A CULTURAL HISTORY OF MATHEMATICS

VOLUME 3

A Cultural History of Mathematics
General Editors: David E. Rowe and Joseph W. Dauben

Volume 1
A Cultural History of Mathematics in Antiquity
Edited by Michael N. Fried

Volume 2
A Cultural History of Mathematics in the Medieval Age
Edited by Joseph W. Dauben, Clemency Montelle, and Kim Plofker

Volume 3
A Cultural History of Mathematics in the Early Modern Age
Edited by Jeanne Peiffer and Volker Remmert

Volume 4
A Cultural History of Mathematics in the Eighteenth Century
Edited by Maarten Bullynck

Volume 5
A Cultural History of Mathematics in the Nineteenth Century
Edited by Tom Archibald and David E. Rowe

Volume 6
A Cultural History of Mathematics in the Modern Age
Edited by Tom Archibald and David E. Rowe

A CULTURAL HISTORY OF MATHEMATICS

IN THE EARLY MODERN AGE

VOLUME 3

Edited by Jeanne Peiffer and Volker Remmert

BLOOMSBURY ACADEMIC
LONDON • NEW YORK • OXFORD • NEW DELHI • SYDNEY

BLOOMSBURY ACADEMIC
Bloomsbury Publishing Plc
50 Bedford Square, London, WC1B 3DP, UK
1385 Broadway, New York, NY 10018, USA
29 Earlsfort Terrace, Dublin 2, Ireland

BLOOMSBURY, BLOOMSBURY ACADEMIC and the Diana logo are trademarks of Bloomsbury Publishing Plc

First published in Great Britain 2024

Series design by Raven Design
Cover image: Portrait of the Mathematician Fra Luca Pacioli and his student by Jacopo De Barbari. (1500–1525) Capodimonte National Museum and Galleries. Naples © Photo courtesy Getty Images

A catalogue record for this book is available from the British Library.

A catalog record for this book is available from the Library of Congress.

ISBN: HB: 978-13500-6295-5
ePUB: 978-13502-9482-0
ePDF: 978-13502-9481-3
Set: 978-13500-6312-9

Series: The Cultural Histories Series

Typeset by RefineCatch Limited, Bungay, Suffolk
Printed and bound in Great Britain

CONTENTS

List of Illustrations vii

List of Contributors xii

Series Preface xv

Introduction 1
Jeanne Peiffer and Volker Remmert

1 Everyday Numeracy: Everyday Numeracy through the Mirror of Textbooks 37
Maryvonne Spiesser

2 Practice and Profession: Range and Change in Mathematical Lives 67
Jim Bennett

3 Inventing Mathematics: Betwixt Tradition and Innovation 97
Sébastien Maronne

4 Mathematics and Worldviews: Mathematics as a Touchstone for "Certainty" 123
David Rabouin

5 Describing and Understanding the World: Setting up the Empire of Mathematics 147
Antoni Malet

6 Mathematics and Technological Change: The Silent Rise of Practical Mathematics 179
Thomas Morel

7 Representing Mathematics: Promoting the Mathematical Sciences in the Early Modern Period 207
Robert Goulding and Volker Remmert

Bibliography 235

Index 245

ILLUSTRATIONS

FIGURES

Introduction

0.1 *Académie de l'espée* (Sword Academy) by Girard Thibault from Antwerp, [Leiden], 1628, Tab. XII. Purchased with income from the Jacob S. Rogers Fund, The Metropolitan Museum of Art. 19
0.2 The instrument "nocturnal" described by Petrus Apianus in his *Cosmographicus liber* (1524), second to last and last pages. Bayerische Staatsbibliothek München: Rar. 271. 23
0.3 The instrument "nocturnal" described by Petrus Apianus in his *Cosmographicus liber* (1524), second to last and last pages, with two printed pieces to be cut out. Bayerische Staatsbibliothek München: Rar. 271. 24

Chapter 1

1.1 An addition of money (livres, sous, deniers) carried out with counters (Estienne de la Roche 1538, *Arismetique*, n.p.). Toulouse, Bibliothèque d'étude et du patrimoine. Photo by Maryvonne Spiesser. 40
1.2 A *suanpan*. Photo by Maryvonne Spiesser. 41
1.3 A multiplication by gelosia, or "To multiply by a checker table." Photo by Maryvonne Spiesser. 43
1.4 Arithmetica, in Georg Reisch. 1508. *Margarita philosophica cum additionibus novis*. Basel: Schott. Bayerische Staatsbibliothek München: Res/4 Ph.u. 117. 45
1.5 Napier's rods. Photo by Patrick Guenette / Alamy Stock Photo. 47

1.6 Alexandre Jean 1649. *L'arithmétique au miroir . . .*: 10, kept in Toulouse, Bibliothèque d'étude et du patrimoine. Photo by Maryvonne Spiesser. 63
1.7 A table of expenses and wages, *Syderum Secreta* (1689), by John Harrison, on Dr. Katherine Walker's website: https://www.katherine-walker.com/research (accessed on April 26, 2023). Scan by Katherine Walker from her personal collection. 64

Chapter 2

2.1 Printed portrait of Gemma Frisius, by Jan van Stalburch, 1557. Courtesy of Rijksmuseum. 69
2.2 Printed portrait of Christopher Clavius, by Francesco Villamena, 1606. Photo by Heritage Art / Heritage Images / Getty Images. 78
2.3 Title-page from Aaron Rathborne, *The Surveyor*, 1616. Photo by Science & Society Picture Library / SSPL / Getty Images. 86

Chapter 3

3.1 Apollonii Pergaei Conicorvm Libri Qvattvor. Bononiae: Benatius, 1566. Courtesy of Max Planck Institute for the History of Science, Berlin. 99
3.2 *AB* and *BC* are the abscissa and the ordinate of point C. Descartes's solution of Pappus's locus problem in *Discours de la méthode pour bien conduire sa raison, & chercher la verité dans les sciences. Plus La dioptrique, les météores et la géométrie, qui sont des essais de cete méthode* (1637: 325). Bayerische Staatsbibliothek München: Rar. 1849. 105
3.3 The rule for solving the equation $x^3 + 6x = 20$ in Cardano, *Great Art* (1545), Chap. XI, p. 30 *recto*. The solution given is $\sqrt[3]{\sqrt{108}+10}-\sqrt[3]{\sqrt{108}-10}$. Bayerische Staatsbibliothek München: 2 Math, p. 8. 109
3.4 Torricelli, "De Dimensione Parabolae," *Opera Geometrica* (1644: 95). Max Planck Institute for the History of Science, Berlin. 115
3.5 The differentiation rules in Leibniz, "Nova Methodus pro Maximis et Minimis, itemque tangentibus," *Acta eruditorum* (1684: 467). Niedersächsische Staats- und Universitätsbibliothek Göttingen. 120

Chapter 4

4.1 The Latin translation of Proclus's commentary on Euclid by Francesco Barozzi: *Procli Diadochi Lycii In primum Euclidis* (1560). Courtesy of Internet Archive. 129

4.2 A Lullian combinatorics of concepts in Athanasius Kircher, *Ars Magna Sciendi, in XII Libros Digesta*, Amsterdam, J. Jansson and E. Weyerstraet (1669: 8). Herzog August Bibliothek Wolfenbüttel: A: 6.3 Quod. 2°. 137

4.3 A diagram for the *Yi-Jing* owned by Leibniz (Letter from Bouvet, 1701. Gottfried Wilhelm Leibniz Bibliothek – Niedersächsische Landesbibliothek, Hannover Signatur: LBr. 105, fol. 27r. 144

Chapter 5

5.1 Giovanni Battista Riccioli, *Almagestum novum astronomiam veterem novamque complectens* (Bologna, 1651), frontispiece with the Copernican and semi-Tychonic systems. Herzog August Bibliothek Wolfenbüttel: A: 2.1.3 Astron. 2° (1). 149

5.2 The painting plane intersecting the visual pyramid in Vignola's *Le due regole della prospettiva pratica* (p. 55). © Universitätsbibliothek Heidelberg. 157

5.3 A simplified method to draw the image of a chequered square lying on the floor, from *Vignola's Le due regole della prospettiva pratica* (p. 37). Max Planck Institute for the History of Science, Berlin. 158

5.4 The geometrical figures accompanying Galileo's proof of the mean-speed theorem, from his *Discorsi e dimostrazioni matematiche* (1638: 170–1). Bayerische Staatsbibliothek München: Rar. 4085. 166

5.5 The geometrical figure accompanying Galileo's proof of the parabolic shot from his *Discorsi e dimostrazioni matematiche* (1638: 242). Bayerische Staatsbibliothek München: Rar. 4085. 167

5.6 Huygens' pendulum clock (from *Horologium oscillatorium* (1673: 4). The cheeks are reproduced under the caption "FIG. II". They are represented sidewise in "FIG.I", marked T. Photo by Photo 12 / Universal Images Group / Getty Images. 170

5.7 From *Horologium oscillatorium* (1673: 66). Gottfried Wilhelm Leibniz Bibliothek. 171

Chapter 6

6.1 Jan Adriaensz. Leeghwater, succession of three waterwheels (powered by windmills) to draw the water from the lowest to the highest level (copy made in 1633). Noord-Hollands Archief: IALW A(492.631.010)029. 185
6.2 Map of the Beemster lake, 1607. Regionaal Archief Alkmaar: PR 1004144. 188
6.3 Plan of a Saxon mine by Johann Scherez, De Geometria Subterranea. Universitätsbibliothek Georg Agricola: WA XVII 15. 192
6.4 Drawing of a gauging rod. ETH-Bibliothek Zürich, Rar 583. 196
6.5 How to measure a cask. ETH-Bibliothek Zürich, Rar 583. 197
6.6 The fortress of Nicosia in Giuseppe Rosaccio, *Viaggio da Venetia, a Costantinopoli per mare, e per terra*, Venice, Giacomo Franco, 1598. Courtesy of Laskaridis Foundation / Europeana. 199
6.7 Seven discarded machine designs for moving the obelisk, from Domenico Fontana, *Della trasportatione dell'obelisco vaticano*, 1590, p. 8. Bibliothèque de Institut national d'histoire de l'art: NUM FOL RES 126. 202

Chapter 7

7.1 Adam and Seth, in: *Chronica regia coloniensis*. Herzog August Bibliothek Wolfenbüttel: GS 04-0515. 209
7.2 Scaliger: *Cyclometrica elementa* duo, p. 87. Max Planck Institute for the History of Science, Berlin. 217
7.3 Three men read while a women writes numerals; representing arithmetic. Engraving by C. Cort, 1565, after F. Floris, *c.* 1557. Wellcome Collection. Public Domain Mark. 219
7.4 Faulhaber, Johannes. 1630. *Ingenieurs-Schul / Erster Theyl.* Herzog August Bibliothek Wolfenbüttel: A: 26.1 Bell. (1). 220
7.5 Tartaglia, Niccolò: *Nova scientia*. Vinegia: S. da Sabio. Courtesy of the Smithsonian Libraries. 221
7.6 Cesare Ripa: *Iconologia.* © Universitätsbibliothek Heidelberg. 223
7.7 *Euclidis quae supersunt omnia*, David Gregory (ed.). Courtesy of The Linda Hall Library of Science, Engineering & Technology. 224
7.8 Jacques Ozanam: *Dictionnaire mathématique*. Herzog August Bibliothek Wolfenbüttel: A: 7.2 Geom. 228
7.9 Jonas Moore: *A New Systeme of the Mathematicks*, London 1681. ETH-Bibliothek Zürich: Rar 5492. 230
7.10 Galileo Galilei: *Il Saggiatore*, Rome 1623. Photo by SSPL / Getty Images. 232

DIAGRAMS

Chapter 1

1.1 How to represent 287,965 “on the lines”. From Recorde 1543, *The Ground of Artes*: 118. Courtesy of Maryvonne Spiesser. 40
1.2 A multiplication performed with Napier’s rods: 456 × 38. Photo by Maryvonne Spiesser. 48

Chapter 3

3.1 Barrow’s method applied to the tangent of a parabola. Courtesy of Sébastien Maronne. 117

Chapter 4

4.1 A diagram for Euclid’s *Elements* I.32. Courtesy of David Rabouin. 130

Chapter 5

5.1 The planet, M, uniformly rotates on its epicycle around center P, which describes a circle (the deferent) centered in D, near the earth T. P moves with uniform angular velocity around the equant, E, located on the line joining D to the earth so that ED = DT. Courtesy of Antoni Malet. 149
5.2 From Kepler’s *Paralipomena* (p. 188), simplified. Courtesy of Antoni Malet. 161
5.3 From Kepler’s *Paralipomena* (p. 196), arrows added. Courtesy of Antoni Malet. 162

CONTRIBUTORS

Jim Bennett, who died on 28 October 2023, spent his working life in university and national museums. This included being Director of the Museum of the History of Science, University of Oxford, and Visiting Keeper at the Science Museum, London, UK. His publications include, "Adventures with Instruments: Science and Seafaring in the Precarious Career of Christopher Middleton," *Notes and Records: The Royal Society Journal of the History of Science* (2019); and *Museo Galileo: Catalogue of Surveying and Related Instruments* (2022).

Robert Goulding is the Director of the Graduate Program in History and Philosophy of Science at the University of Notre Dame, IN, USA, and an Associate Professor of Liberal Studies. His current projects concern Neoplatonic natural philosophy, and the history of the law of refraction. Two of his most recent publications include, "Geometry and the Gods: Theurgy in Proclus's *Commentary on the First Book of Euclid's Elements*," *Perspectives on Science* (2022); and "The Harvest of Optics: Descartes, Mydorge, and Their Paths to a Theory of Refraction," *Annals of Science* (2022).

Antoni Malet has been Full Professor of History of Science at the Universitat Pompeu Fabra, Barcelona, and is now affiliated with the Institut d'Història de la Ciència, Universitat Autònoma de Barcelona, Spain, and associated with the Laboratoire SPHERE (Sciences, Philosphie, Histoire), Centre national de la recherche scientifique (CNRS), Paris, France. He has written on sixteenth- and seventeenth-century mathematics and optics, optical and mathematical instruments, and the early modern philosophy of mathematics and mathematization. Since January 2022, he has been co-editor-in-chief of the journal *Historia Mathematica*. His recent publications include, "Newton's Mathematics," in *The Oxford Handbook of Isaac Newton*, edited by Eric

Schliesser and Chris Smeenk (2020); "Milliet Dechales as Historian of Mathematics," *Perspectives on Science* (2022); and "Some Thoughts on Pascal's and Galileo's 'Indivisibles' and the Infinite Divisibility of Extension," *Revue d'histoire des sciences* (forthcoming, 2023).

Sébastien Maronne is Associate Professor of History and Philosophy of Mathematics at the Mathematical Institute of Toulouse, France. He works on the history of early modern mathematics (Descartes, Roberval, Pascal) and on the French scientific epistemology (Bachelard, Cavaillès, Vuillemin). He is the editor with June Barrow-Green and Niccolò Guicciardini of the special issue of *Historia Mathematica*, "Contexts, Emergence and Issues of Cartesian Geometry: In Honour of Henk Bos's 70th Birthday" (2010). Sébastien's recent publications include, "Descartes's Mathematics," in Steven Nadler, Tad M. Schmaltz, and Delphine Antoine-Mahut (eds.), *The Oxford Handbook of Descartes and Cartesianism* (2019); and "Dettonville et les données," *Revue d'Histoire des Sciences* (forthcoming, 2023).

Thomas Morel received his PhD in the history of science at the University of Bordeaux, France. After working at the Technical University Berlin, Germany, and the Université de Lille, France, he is now Professor of History of Mathematics and Mathematics Education at the Bergische Universität Wuppertal, Germany. His research focuses on mathematical practitioners, useful knowledge, and the social uses of quantification. Among his publications are, "Bringing Euclid into the Mines: Classical Sources and Vernacular Knowledge in the Development of Subterranean Geometry," in *Translating Early Modern Science* (2017); and "De Re Geometrica: Writing, Drawing and Preaching Mathematics in Early Modern Mines," *Isis* (2020). Thomas' last book, *Underground Mathematics* (2023), presents a cultural history of mining geometry in early modern Europe.

Jeanne Peiffer has made her career as a research scientist at the Centre National de la Recherche Scientifique (CNRS), and was affiliated with the Centre Alexandre Koyré, Paris, France, where she is still active as emerita. As a historian of mathematics, Jeanne focused on early modern mathematics, namely on Albrecht Dürer's contributions to geometry, publishing her book *Albrecht Dürer, Géométrie* (1995); and on Johann Bernoulli's correspondence, about which, see *Der Briefwechsel von Johann Bernoulli*, 2–4 (1988, 1992, and forthcoming). Jeanne is also the author of a popular *Histoire des mathématiques* (1982, translated into Russian, German, American English, and Spanish). More recently, she is part of an international network devoted to the study of the circulation of mathematics within and via journals, cf. *Interplay between Journals at Various Scales*, special issue of *Historia Mathematica* (2018). For more details, see: https://cak.ehess.fr/membres/jeanne-peiffer.

David Rabouin is a Senior Research Fellow at the Centre national de la recherche scientifique (CNRS), Paris, France, in the research group SPHERE (Sciences, Philosophie, Histoire, UMR 7219, CNRS, Université Paris Cité). His interest is in the history of philosophy and mathematics in the early modern period, with special focus on Descartes and Leibniz. He is the head of the advanced grant PHILIUMM (The Philosophy of Leibniz in Light of His Unpublished Mathematical Manuscripts) of the European Research Council (Adg n°101020985). He is the author of several books, including, *Mathesis universalis: L'idée de 'mathématique universelle' d'Aristote à Descartes* (2009); *G.W. Leibniz, Interrelations between Mathematics and Philosophy*, with Norma B. Goethe and Philip Beeley (2015); and *G.W. Leibniz: Mathesis universalis: Ecrits sur la mathématiques universelle*, with the "Mathesis group" (2018).

Volker Remmert has been trained as a mathematician and historian. He teaches History of Science and Technology at the University of Wuppertal, Germany. His research interests are in the history of early modern science and in the history of mathematics in Germany in the nineteenth and twentieth centuries. Volker is the author of "Oberwolfach in the French Occupation Zone: 1945 to Early 1950s," *Revue d'histoire des mathématiques* 21(2020); co-author with Julia Ellinghaus of "Visual Worlds on Early Modern Scientific Instruments: Types and Messages," in Ruth Noyes (ed.), *Reassessing Epistemic Images in the Early Modern World* (2023); and co-editor with her of *Manipulating the Sun: Picturing Astronomical Miracles from the Bible in the Early Modern Era* (forthcoming, 2023).

Maryvonne Spiesser was Lecturer in Mathematics and History of Mathematics at the University of Toulouse III, France, until 2010. Her research focuses on the history of practical arithmetic and algebra in Europe between the Middle Ages and the Renaissance, and on seventeenth-century mathematics. Currently she is involved in the popularization of mathematics within the association Fermat Science and in the realization of a center for mathematical culture in Beaumont de Lomagne, Fermat's birthplace in Southwest France. Maryvonne is the author of *Une arithmétique commerciale du XVe siècle: Le* Compendy de la praticque des nombres *de Barthélemy de Romans* (2003). Among her more recent publications are, "Pedro Nunes: Points de vue sur l'algèbre," in Sabine Rommevaux, Maryvonne Spiesser, and Maria Rosa Massa (eds.), *Pluralité de l'algèbre à la Renaissance* (2012); "La *Spéculative des nombres* (*ca.* 1475), un pont entre l'arithmétique pratique et ses fondements théoriques: Analyse et édition du texte," *Revue d'histoire des mathématiques* (2022).

SERIES PREFACE

Culture, history, and mathematics are words that seldom appear together. Yet, mathematical activities have played a pervasive role in diverse cultures throughout human history, as described in the present six volumes. These illustrate how individuals and societies have engaged with various manifestations of mathematical culture from antiquity to the present. Based on recent and ongoing scholarship, they provide a global overview of a highly diverse range of developments, though certainly nothing like a comprehensive picture. Editors and authors had to make difficult choices about which topics and cultures to include, and in doing so, they have sought to strike a balance between characteristic features of a period and particular individuals and events that had a deep influence or longer lasting impact.

Each volume addresses seven general themes in as many chapters. The first three chapters—Everyday Numeracy, Practice and Profession, and Inventing Mathematics—highlight the roles of those who used, taught, or created mathematics as it evolved over time. These chapters explore mathematics per se, whereas the three that follow—Mathematics and Worldviews, Describing and Understanding the World, Mathematics and Technological Change—consider the larger impact of mathematics on other dimensions of human culture. The final chapter—Representing Mathematics—is devoted to a broader understanding of mathematical ideas, activities, or artifacts in different cultures throughout the world. The periods covered are:

Volume 1: A Cultural History of Mathematics in Antiquity
(3000 BCE–500 CE)
Volume 2: A Cultural History of Mathematics in the Medieval Age
(500–1400)

Volume 3: A Cultural History of Mathematics in the Early Modern Age (1400–1687)
Volume 4: A Cultural History of Mathematics in the Eighteenth Century (1687–1800)
Volume 5: A Cultural History of Mathematics in the Nineteenth Century (1800–1914)
Volume 6: A Cultural History of Mathematics in the Modern Age (1914–2020)

General Editors: David E. Rowe and Joseph W. Dauben

Introduction

JEANNE PEIFFER AND VOLKER REMMERT

Most readers of this volume will have a general idea of "mathematics" and so, to start with, we use the term in a contemporary way. Later in this introduction, however, we will explain what mathematics and related notions meant in the early modern period. Delving into the history of early modern mathematics, its practitioners, their contexts, and its many uses, means to engage with a world that is unfamiliar to most people today. This introduction paves the way for the reader of this volume, which bridges several centuries while focusing on the early modern period. The latter extends roughly from the Fall of Constantinople (1453) to the publication of Isaac Newton's *Philosophiae naturalis principia mathematica* (Mathematical Principles of Natural Philosophy, 1687), henceforth *Principia*. It was a time of tumultuous, even revolutionary changes that mark the beginning of an era of increasing interplay between science and society in the Western World.

We will discuss how the wide range of social, cultural, and intellectual changes affected mathematics as well as the ways in which the mathematical sciences affected these changes. The geographical and historical focus of the volume is on Western Europe and its literate traditions, especially in France, Germany, Italy, and Great Britain, where we find important centers of knowledge. Libraries, archives, and museums serve as the main repositories for the abundant source material—books, handwritten texts, letters, and instruments—documenting the development of mathematics in these centers. Unlike the case with Western Europe, historians of mathematics are only beginning to explore systematically the contributions made in other parts of the world. Mathematical knowledge played an especially important role in

Islamicate societies, and a great deal of that knowledge gradually migrated to the West via Spain and Italy, where it nourished the development of the mathematical sciences considerably, particularly in the field of algebra, but also in practical geometry and astronomy.

We open this introduction by giving a brief overview of the political, religious, and cultural changes that took place in the period. This was an era when intellectuals not only looked back to the past, they also went consciously beyond it, partly by modifying older traditions, but also by inaugurating a new approach to the study of nature. The latter approach led to what twentieth-century historians have called the Scientific Revolution, the topic of the second section. Mathematics or the mathematical sciences played a significant role in this revolution, as it progressively became one of the principal tools for describing and understanding the world. In order to describe these mathematical activities, we focus in the third section on those who cultivated mathematics as well as the settings in which those activities took place. We then discuss the material and intellectual resources that mathematicians shared with other cultural actors, but also the specific tools mathematicians put in place. As mathematics gradually pervaded ever larger parts of early modern society, it gained importance as a resource for other fields of inquiry. As a result, the social and epistemological status of mathematics rose considerably, making it the focus of an ongoing debate. Finally, we turn to the new mathematical ideas studied during the period, including shifts in the kinds of questions asked and the methods used to answer them. Throughout this introduction, we provide references to various topics discussed in the seven chapters of this volume. The chapters are concerned, in general, with the seven themes addressed throughout this cultural history of mathematics: (1) numeracy, (2) profession, (3) invention, (4) worldviews, (5) mathematization, (6) technology, and (7) representation.

THE EARLY MODERN PERIOD AS A PERIOD OF CHANGE

The Fall of Constantinople, the capital of the Byzantine Empire, to the Ottoman Turks marks the beginning of a period of dramatic transformation in Western Europe. This event not only spelled the end of the Eastern Roman Empire, but in the eyes of some historians it also brought the Middle Ages to a close and sparked what is known as the Renaissance. The term Renaissance, referring to a cultural rebirth or revival, is somewhat controversial for two main reasons. On the one hand, it focuses almost entirely on Western Europe; on the other, historical studies have clearly shown that significant cultural developments continued to take place throughout the Middle Ages. Protestant historians have tended to identify the birth of modern Europe with Luther's posting of his famous theses on the front door of the church in Wittenberg in 1517, which marks the beginning of the Reformation. The instauration of a new Christian confession led to a period of instability, seditions, and religious wars during the second half of the sixteenth and the first half of the seventeenth centuries. The movable-type printing press, invented by the German Johannes Gutenberg (*c.* 1400–68) in the middle of the fifteenth century, facilitated the spread of Protestantism and the written word in general. Note, however, that woodblock printing was already in use since the eleventh century in ancient China.

During the same period, the voyages of discovery radically altered European understanding of the world at large as well as representations of it. Henceforth, this included a new continent, the "New World" of the Americas, discovered in 1492 by Christopher Columbus (1451–1506) and drawn on a map by Martin Waldseemüller (1470–1520) in 1507. Columbus's exploits as well as the circumnavigations of the world by Ferdinand Magellan (1480–1521) in 1519–21 brought new challenges for pilots, cartographers, and cosmographers. They soon came to recognize the importance of mathematical principles as the basis for navigational and mapping techniques (cf. Chapter 5). Encounters with the "New World" and with hitherto unknown peoples, their traditions and cultures could be enriching and stimulating; they were, however, anything but straightforward. Such experiences often required negotiations, for which seamen, merchants, and missionaries were little prepared. The English nobleman and chemist Robert Boyle (1627–91) offered instructions for travelers and navigators in his *General Heads for the Natural History of a Country* (1692), in which he described strategies for gathering knowledge. Over the course of the period, early trade links between Italy and the Ottoman Empire gradually expanded to form a global network of commercial enterprises. These businesses introduced double-entry bookkeeping in their financial transactions (cf. Chapter 1). Their ships sailed the oceans, bringing lucrative goods, such as spices and silks from the East, to Europe. Modern financial and insurance systems developed along with an international banking system centered in Italy.

The rich exchanges with other civilizations in the context of European expansion and the circulation of goods and luxury commodities enhanced the quality of urban life, which rose to a level never attained before. A prosperous merchant class emerged and, along with it, the spread of numeracy (cf. Chapter 1). Important innovations in intellectual life also took place. Thus, among the émigrés who fled Constantinople to find refuge in the Christian West, and especially in the Italian city states, were several high-profile scholars who brought with them the legacy of Greek culture and knowledge. Most of them settled in Venice, but they also resided in Florence, Rome, Naples, and other cities, where they taught Greek, translated texts into Latin, or copied manuscripts for their learned patrons. They thus contributed a great deal to the recovery of ancient Greek knowledge.

In 1494, Aldus Manutius (1452–1515), a former university teacher, founded the printing house Aldine Press in Venice. His company employed otherwise unknown Greek émigrés to produce valuable printed editions of a large number of works from Greek antiquity. Manutius' effort was part of the humanist movement, which originated in Florence during the fourteenth, spread throughout Italy in the fifteenth, and exerted a massive influence on European culture by the sixteenth century. The recovery of Plato's original writings enriched the philosophical literature of the time (cf. Chapter 4), just as mathematical texts played a major part in these humanistic printing endeavors. A Latin translation of Euclid's *Elements* was already published in 1482, followed by a Latin edition of his collected works in 1505 (reprinted in 1510, 1516, 1537, 1546, and 1558), and in 1533 a Greek text of the *Elements* was published by Simon Grynaeus (1493–1541) in Basel. In 1537, a Latin translation of Apollonius's works was put on the market in Venice, including the first four of his eight books on conic sections (*Conica*). A Greek and Latin Archimedes edition came out in 1544, while Pappus's *Mathematical Collection* only appeared in 1588. Thus, the availability of classical mathematical texts quickly spread throughout Latin Europe (Rose 1975).

Humanism promoted the educational goals of aspiring non-clerical elites in an urban environment. The aims of humanist learning were primarily directed at a good command of Latin and Greek, based on the reading of classical sources. Originating in the Ciceronian rhetorical tradition, humanism included areas of knowledge such as grammar, rhetoric, poetry, moral philosophy, and ancient history, but its influence quickly expanded to other areas, including mathematics. During the early modern period, various traditions came together as accepted parts of general intellectual culture. Besides the classical Greek tradition, often mixed with Byzantine or Arabic elements, there was also a strong emphasis on the medieval Latin tradition, which drew on the works of Aristotle (cf. Chapter 4). This merging of a variety of traditions acted as a cradle for a dynamic understanding of knowledge that was susceptible to new developments. Scholars claimed that a new epoch had begun and proudly highlighted their inventions and discoveries in book titles pointing to a new method, a new science, etc. Newness was no longer a disturbing feature; many saw it positively, as embracing innovation became one of the distinguishing features of the early modern period.

THE "SCIENTIFIC REVOLUTION"

From the perspective of the history of science, the most prominent development of the early modern period was the so-called Scientific Revolution of the sixteenth and seventeenth centuries, a concept coined in 1939 by the historian of science Alexandre Koyré (1892–1964) (Dear 2001; Osler 2010). It gave rise to new approaches to the study of nature, and these led to new ideas and knowledge, which had close links with many of the changes discussed above. One of the fundamental features of the Scientific Revolution was a significant extension of the realm of experience. This went along with a growing appreciation of practical knowledge in general as well as scientific instruments, in particular (e.g., microscopes, telescopes, air pumps, barometers, etc.). Both aspects played a key role in approaching the study and understanding of nature by way of experimentation and systematic observation. Moreover, this empirical approach went hand in glove with processes of mathematization, as over the course of the period mathematics emerged as the dominant tool not only for describing, but also for explaining the world and its many phenomena. Leading natural philosophers thus came to regard the language of mathematics as a royal road to knowledge about the world, in stark contrast to earlier periods. While awareness of the terrestrial world thus expanded through the voyages of discovery, knowledge of the celestial world broadened vastly soon thereafter, as Koyré famously described in *From the Closed World to the Infinite Universe* (1957).

As a long-term process, the Scientific Revolution unfolded from the mid-sixteenth to the late seventeenth century as a period of previously unimaginable vitality in the sciences. Two famous works mark its beginning and end. At the outset, the publication of Copernicus's epoch-making astronomical text *De revolutionibus orbium coelestium* (On the Revolutions of the Heavenly Spheres, Nuremberg 1543) placed the sun at the center of the then known universe, thereby making the earth a planet in our solar system. The triumph of this heliocentric system came in 1687 with the appearance of Newton's *Principia*, which introduced the universal law of gravitation to account for the motions of all bodies in the heavens and on earth. In broad terms, the Scientific Revolution permeated all branches of knowledge about nature and the world—from anatomy, biology, and natural history to astronomy, cosmology and physics—doing away with old certainties and embarking on new conceptual and methodological roads. In his poem *Anatomy of the World* (*c.* 1612), the English poet John Donne (1572–1631) described the shockwaves brought forth by what we now call the Scientific Revolution, and in particular by the Copernican displacement of the earth from the center of the universe:

> And new philosophy calls all in doubt,
> The element of fire is quite put out;
> The sun is lost, and th' earth, and no man's wit
> Can well direct him where to look for it.
> And freely men confess that this world's spent,
> When in the planets, and the firmament
> They seek so many new; they see that this

Is crumbled out again to his atomies.
'Tis all in pieces, all coherence gone;
All just supply, and all relation.

Five iconic figures in the Scientific Revolution, whose ideas and results shattered older notions and worldviews, pursued approaches based solidly on mathematics. Copernicus drew on earlier works in mathematical astronomy in proposing a new heliocentric world system. Galileo Galilei (1564–1642) founded mechanics and the law of free fall, placing both in the context of a geometrical tradition dating back to Archimedes and Apollonius. Kepler, too, drew inspiration from the works of Greek geometers in setting forth his laws of planetary motion, which led to a further mathematization of the heavens. After René Descartes (1596–1650) introduced the powerful tool of analytic geometry, Newton unified terrestrial and celestial mechanics by means of his laws of motion and the law of universal gravitation.

All of these thinkers lived in a fast-changing world, in which new communication structures greatly facilitated their own work, as correspondence networks became increasingly important. One of the most important information brokers in Paris was the Minim monk Marin Mersenne (1588–1648), who acted as the central node in an international network of scientific contacts. Mersenne corresponded regularly with, among others, René Descartes, Pierre de Fermat (1607—65), Galileo Galilei, and Blaise Pascal (1623–62), and kept them up-to-date about new ideas, publications, or experiments. In England, the German immigrant Henry Oldenburg (*c.* 1618–77) maintained an enormous correspondence network, which he cultivated further after he became secretary of the newly founded Royal Society of London. Oldenburg circulated excerpts of the many letters he received, which in 1665 gave rise to one of the first printed journals in the history of science, the *Philosophical Transactions* of the Royal Society, which began publication in the same year as the Parisian *Journal des Sçavans*. The Jesuits were a global intellectual elite. In their head offices in Rome, they amassed reports throughout the sixteenth and seventeenth centuries written by Jesuit authors drawing on information gathered at their missions from all over the world (China, India, and the Americas) and covering all branches of knowledge. This worldwide knowledge flow largely ignored confessional or religious boundaries.

New forms of knowledge were discussed at newly founded scientific academies, such as the Accademia dei Lincei in Rome (1603), the Accademia del Cimento in Florence (1657–67), the Royal Society of London (1660) or the Académie des Sciences in Paris (1666). Less formal discussions took place elsewhere: in coffeehouses and at courts, in colleges and workshops, on ships and while traveling. Printing allowed for a wide and relatively easy dissemination of knowledge and with hitherto unknown speed, facilitated by the use of Latin as an academic language. Since Europeans regarded their knowledge, particularly in mathematics, to be universal, they also used it in their missionary activities. The Jesuit Matteo Ricci (1552–1610), a key figure in the Jesuit mission to China, offers a prime example. He instigated a Chinese translation of the Latin edition of Euclid's *Elements* prepared by his fellow Jesuit Christopher Clavius (1538–1612) as part of his Jesuit missionary enterprise (cf. Chapter 2; Hsia 2009; Jami 2011). In a similar way, the *Typographia Medicea* in Rome (Medici Oriental Press, 1584–1614), which specialized in producing Arabic texts, printed 1,500 copies of the *Gospels* (1590/91) and 3,000 copies of Euclid's *Elements* (1594) in Arabic, both for missionary purposes; the latter slightly outsold the former (Saliba 2007: 228f.).

Narratives of the Scientific Revolution have often stressed two central characteristics. First, this was a dynamic period, which saw extraordinary developments in the sciences, and these gave rise to countless myths about progress. Second, it became a milestone in European and World history because it marks the beginning of the scientization of modern societies. This scientization is characterized by a close and growing correlation as well as mutual exchange between scientific and technological developments, on the one hand, and social and cultural changes, on the other. It should, however, not be forgotten that these processes were deeply rooted in earlier developments in the Islamicate and Latin Worlds. For instance, the astronomical work of Copernicus drew on the writings of a number of authors from Islamicate societies (Swerdlow and Neugebauer 1984; Saliba 2007). Moreover, historians have long ago documented the importance of both the Greek and Islamicate traditions for the development of early modern thought in general. Many of the protagonists of the Scientific Revolution openly acknowledged these traditions and clearly saw themselves as standing on the shoulders of giants.

Nevertheless, and this is part of any account of the Scientific Revolution, many of its leading actors clearly saw that they were embarking on something fundamentally new, not only in terms of how to generate and codify new knowledge, but even in sketching out new visions for the future. In this respect, two figures stand out. The first was the English philosopher, Francis Bacon (1561–1626), often called the "father of empiricism," who coined the phrase *scientia est potentia* (knowledge is power) and inspired both the Royal Society as well as the early Académie des Sciences. Meanwhile, the Florentine Galileo Galilei famously defended Copernicanism, while promoting a complementary vision of scientific progress. Whereas Bacon deeply believed in the potential of scientization, Galileo emphasized the power of mathematization, though both shared a vision of the positive future influence of science and mathematics.

In his *Nova Atlantis* (New Atlantis, 1624), Bacon described a utopian laboratory, the House of Solomon, which later served as a kind of model for the Royal Society. Here, workers systematically gathered and ordered the new and useful with a clear goal in mind: "The end of our foundation [House of Solomon] is the knowledge of causes, and secret motions of things; and the enlarging of the bounds of human empire, to the effecting of all things possible" (Bruce 2009: 177). Bacon's vision for how the sciences could bestow their blessings on society made no place, however, for the mathematical sciences (cf. Chapter 7). Yet, at about the same time, Galileo in his *Il Saggiatore* (The Assayer, 1623), professed his credo of mathematization (cf. Chapter 4). Here, he expounded the oft-quoted notion that the book of nature "is written in the language of mathematics, and its characters are triangles, circles, and other geometric figures without which it is humanly impossible to understand a single word of it" (Drake 1957: 237f). With this, Galileo, in as high a tone as Bacon, made an enormous promise, namely to extract nature's secrets and potentials by means of mathematics and the mathematical sciences.

Mathematization—loosely understood as the potential to describe and analyze the world with mathematical methods (as described in Chapter 5 and below)—is a pervasive concept in the history and historiography of early modern mathematics, often taking on mythic dimensions. As such, one should handle this notion with care. As the historian of mathematics and science, Domenico Bertoloni Meli, nicely expressed this:

> [it] is one thing to state a priori, following a tradition loosely derived from Plato that the motion of a cannonball or the flow of water in a river can be described mathematically, but it is quite another actually to produce a meaningful quantitative description of such complex phenomena. (Bertoloni Meli 2006: 10)

EARLY MODERN MATHEMATICS/ MATHEMATICAL SCIENCES

To understand what mathematics in the sixteenth and seventeenth centuries meant, we now turn to discuss the categories used by the actors when they talked about mathematics, but also the activities, which they or their contemporaries labeled as mathematical. Let us first look at the texts written during the early modern period, mainly by professors teaching mathematics at universities and colleges. For them, the term mathematics, or the then more usual "mathematical sciences" (*scientiae* or *disciplinae mathematicae*), designated a broad range of fields that encompassed not only arithmetic and geometry but also domains such as astrology and astronomy, music, optics, geography, navigation, architecture, surveying, etc. They generally introduced a subdivision to distinguish between pure and mixed mathematics. Pure mathematics dealt with quantities, which could be discrete, like numbers in arithmetic, or continuous magnitudes, i.e. lengths, areas or volumes, as studied in geometry. Mixed mathematics dealt not only with quantities but also with qualities, among them the aforementioned fields of knowledge as well as many others, according to the enumerations presented by various authors. The Ulm mathematician and fortress architect, Johannes Faulhaber (1580–1635), for instance, presented twelve classifications of specific branches of the mathematical sciences on the engraved title page of his *Ingenieurs-Schul* (School for Engineers) from 1630 (cf. Chapter 7). Somewhat later, the Jesuit Gaspar Schott (1608–66) enumerated almost 30 specific fields within the mathematical sciences in his *Cursus Mathematicus* of 1661.

In view of this broad understanding of mathematics, it comes as no surprise that mathematicians engaged in a wide number of activities beyond computing and measuring. These ranged from time keeping to mapmaking and from surveying to architecture, including instrument making, engineering, astronomy, casting of horoscopes, etc. The mathematical sciences thus operated at different levels of sophistication between theoretical abstraction and practical application, and they encompassed various fields of knowledge, many of which do not fall under the heading of mathematics today. The large variety of figures portrayed in Chapter 2, described as "an oddly miscellaneous and eccentric collection of mathematicians" corroborates the wide and inclusive meaning of the term mathematics as then understood. Those called *mathematician* (or *mathematicus* in Latin) by their sixteenth-century contemporaries could be anyone engaged in designing and making scientific instruments, clocks or globes. They were often involved in a multitude of tasks, whether as cosmographers or astronomers, engravers, printers, cartographers, surveyors, or seamen. Above all, they worked as mathematical practitioners, sharing a hands-on approach informed by mathematical and technical knowledge in order to tackle practical problems. In general, as historian of science Federica Favino put it, "early modern practical mathematicians were a group of professionals who are very difficult to document" (Favino 2017: 710).

A quite large group of people taught the mathematical sciences, as well as practical mathematics, in various contexts and institutions. These included universities and colleges but also various schools in urban contexts, such as abacus schools, naval academies, mining schools, as well as private households. Those who taught mathematics presumably tried to adjust their subject matter to meet the needs of different audiences. At a basic level, their pupils learned methods of counting and computation (the four elementary arithmetic operations) with hopes of attaining a sufficient level of numeracy (cf. Chapter 1). In the urban contexts of merchants and artisans, abacus schools taught sophisticated arithmetical methods (basic algebra, accounting, rule of three). Shipyards and arsenals offered formal apprenticeships to those who wished to learn surveying, gauging, navigation, and mapmaking. Colleges and universities taught the more theoretical branches of the mathematical sciences, the main topics being the four first books of Euclid's *Elements* and elementary astronomy. The standard text for the latter subject was the *Sphaera*, authored by Johannes de Sacrobosco (John of Holywood, *c.* 1195–1256), which gave a simplified treatment of the Ptolemaic universe. Both of these texts went through multiple editions and adaptations (hundreds in the case of Sacrobosco). The teaching of the mathematical sciences was constantly adapted to new audiences, such as young noblemen and aspiring military officers keen to learn about fortification, ballistics, or military strategies. Many of those teaching mathematics in these contexts as well as practical mathematicians were also active as publishing authors.

Although they often shared interests and practices, there were no typical career trajectories for mathematicians in the early modern period. A telling example was Johannes Müller (1436–76), known mainly by his Latinized name, Regiomontanus. As a remarkable prodigy, Regiomontanus entered the University of Leipzig at the age of eleven, but went on to Vienna, where he studied under the astronomer Georg of Peurbach (1423–61), with whom he came to collaborate intensively. There he met Cardinal Bessarion (1403–72), a papal legate, who came to Vienna in order to drum up support against the threat of a Turkish invasion of Latin Europe. A native Greek-speaker, Bessarion promoted classical Greek learning and encouraged Peurbach to prepare a more accurate Latin translation of Ptolemy's *Almagest*, the famous text that detailed the mathematical (geocentric) theory of the motions of the sun, the moon and the planets (cf. Chapter 5).

After Peurbach's death, Regiomontanus completed this work and published the *Epitome in Ptolemaei Almagestum* (Introduction to Ptolemy's Almagest, 1462). Modern scholars regard it as marking a shift from uncritical reverence for Ptolemy and the ancients to a respectful and innovative approach. Regiomontanus dedicated the book to Bessarion, his patron in Rome, where he lived in his extended household and had access to its rich collection of ancient manuscripts. Regiomontanus traveled in Italy, purchasing or copying rare books for his collection, while lecturing on the mathematical sciences. At the University of Padua, he lectured on the *Elements of Astronomy* of al-Farghani, a ninth-century astronomer in Baghdad, known by the Latin name Alfraganus. This was an extensive summary of Ptolemy's *Almagest* containing additional and revised data from the Islamicate tradition; its Latin translation was very popular in Europe at that time. Dante Alighieri (*c.* 1265–1321), the famous author of the *Divina Commedia* (Divine Comedy), appears to have drawn his knowledge of Ptolemaic astronomy from his reading of al-Farghani, and Copernicus may have used the latter's calculation of the diameter of the earth.

From Italy, Regiomontanus went to Buda in Hungary, following his appointment to the Royal Library with its rich collection of ancient manuscripts, part of the booty the king had brought from war with the Turks. There, he came across an incomplete copy of Diophantus' *Arithmetics*, an encounter which enabled this text to become known in Europe (cf. Chapter 3). While in Hungary, he computed tables of sines and finished his treatise on trigonometry, *On Triangles* (1462–4), published only in 1533. He deemed this work necessary to "reach a satisfying knowledge of the stars," as he put it in the preface. In 1471, he settled in Nuremberg, where he made astronomical observations and set up a workshop for manufacturing instruments. He also established a printing press in Nuremberg that published his own works as well as carefully edited ancient and medieval mathematical texts with commentaries—this, not even two decades after Gutenberg first invented printing with movable type. His press issued calendars and almanacs, among them *Ephemerides* (1474) for the years 1475–1506, on which seamen, including Christopher Columbus, relied for navigation. Regiomontanus was a prolific compiler of tables, also for astrological use, notably in medicine. He cast horoscopes for famous patrons and designed a method of finding times for bloodletting according to the position of the moon. In 1475, the pope called him to Rome to work on calendrical reform; he died there at the age of only forty, presumably from an outbreak of the plague.

In condensed form, Regiomontanus' biography points to a number of aspects and activities characteristic for early modern mathematics. Many of its leading figures belonged to the humanist tradition that collected and studied learned books; some engaged in the difficult process of retrieving ancient mathematical texts and making these available through print editions. Many of these were Arabic manuscripts, which made these Europeans keenly aware of Islamicate authors. Thus, in the Latin West, one rarely referred to Claudius Ptolemy's classic astronomical text by its original Greek title; mathematicians called it the *Almagest* from its Arabic name. Like Regiomontanus, many mathematicians of this time built scientific instruments; they typically depended on patronage; many traveled to various centers of learning, lecturing at universities or supplementing their incomes by practicing astrology, while coupling their horoscopes with medical treatments. Of course, the understanding of mathematics and the mathematical sciences underwent important shifts between the fifteenth and the seventeenth centuries, some of them described in this volume, but the vast variety of mathematical cultures was an overall characteristic of the period. As the historian of science Richard J. Oosterhoff noted, "contemporaries always heard with the word 'mathematics' a heterogeneous constellation of connotations and practices" (2018: 6).

VARIOUS SETTINGS FOR MATHEMATICAL PRACTICE

Like all human activity, mathematical practice is located in time and place. The practices of Jesuits at the Beijing Imperial Observatory or those of German university lecturers were not the same, although they probably had much to share and to exchange. As we have seen, Regiomontanus established himself in Nuremberg because of the intellectual and material resources offered by the city, namely its renowned instrument makers, but also because of its favorable geographical location. After he died in Rome, his collection of books, manuscripts and instruments remained in Nuremberg where his disciple and patron, Bernhard Walther (1430–1504), continued to make use of his observatory and library. Nuremberg's Albrecht Dürer (1471–1528) was not only a famous painter and engraver but also the author of three books, one a text on geometry aimed at artists and craftsmen. He had the opportunity to purchase mathematics books from Regiomontanus' library and he studied these with humanist friends better versed in Latin and Greek. This illustrates how particular events in special settings can sometimes have long lasting effects on intellectual practices, including mathematical ones.

If we consider institutional settings where mathematical activity took place, among the first that come to mind are the universities. Indeed, during the early modern period, these medieval institutions remained one of the main sites for learning, while supporting many outstanding figures of the Scientific Revolution. Although universities were unable to cope with the challenges posed by the technical developments in rapidly changing societies, and were often openly hostile to the new learning, they nevertheless still played a certain role in the Scientific Revolution. The curriculum of Renaissance universities stemmed from the medieval universities, which closely followed the traditional models of Bologna and Paris. These had four faculties, but the arts faculty played a propaedeutical role for the three higher faculties of theology, law, and medicine, which students could (theoretically) only enter after having mastered the arts curriculum and obtained the degree of *Magister Artium* (Master of Arts). At the core of the curriculum of the arts faculty were the seven liberal arts (*artes liberales*), divided into three preparatory subjects, the *trivium* (cf. the expression "trivial"), comprised of rhetoric, logic, and grammar, and the *quadrivium*, consisting of the four mathematical subjects: geometry, arithmetic, astronomy, and music.

From the sixteenth century onward, however, general education as provided by the arts faculty acquired a certain independence within the colleges of humanities, sometimes without any links to the university but in other cases incorporated within it. In the former case, lecturers often taught philosophy, mathematics, and the sciences at a higher level. The teaching of arts could lead to a scientific curriculum as, for instance, in Leiden or to a mathematical curriculum as at the Jesuit Roman College. During the sixteenth century, the center of gravity shifted gradually from Bologna and Paris to Padua and Leiden, although the academic map continually changed, as when the Spanish university in Salamanca became especially important for the colonies in the New World. Concerning mathematics, sixteenth-century Italian universities founded specialized chairs, as did the Collège royal, which King Francis I established in Paris in 1530 as an alternative institution to the university. In the seventeenth century, the Scottish universities surpassed Oxford and Cambridge, even though Oxford gained the Savilian Chair of Geometry and Astronomy in 1619 and Cambridge the Lucasian Chair of Mathematics in 1663 (cf. Chapter 2).

Humanistic influences pervaded early modern universities, which adapted to changing times and new audiences. Whereas medieval universities attracted intellectual elites, who were mainly interested in acquiring knowledge for its own sake, humanistic universities trained new elites pursuing careers in state affairs—civil servants, physicians, lawyers, and other professionals. Mathematics was considered useful and thus became an accepted part of the arts curriculum, although in practice it was neglected at many universities. Philip Melanchthon (1497–1560), professor of ancient Greek at the University of Wittenberg (founded in 1502) and Martin Luther's brother in arms, strongly advocated the teaching of the mathematical sciences and gave them a central place in reformed Protestant universities.

Sources documenting the teaching and learning of mathematics at early modern universities are quite plentiful; some are in the form of handwritten notes as well as printed textbooks. The margins of the latter often contain handwritten remarks and commentaries, which often give hints as to who read these books and how they learned mathematics from them. Mathematics was part of the philosophy course, usually taught by members of the arts faculty with limited training in mathematics. On a higher level, astronomical observatories were traditionally important places where mathematical knowledge was gathered, developed, and used. These operated across Europe and Asia, in China, Europe and the Islamicate world. In the latter, the teaching of mathematical knowledge often took place in religious institutions such as mosques or madrasas.

The early modern period also witnessed the creation of new venues dedicated to making and transmitting knowledge, such as botanical gardens, anatomical theaters, cabinets of curiosities, museums, instrument makers' shops, and academies. Printing houses also offered space for scholarly work and debate, as exemplified by the life of Regiomontanus. One might think of the highly educated humanist scholar, Desiderius Erasmus (1466–1536), who lived for years in the house of the printer and publisher Johann Froben (1460–1527) in Basel, one of the leading centers in the European book trade. In general, private dwellings might house traveling scholars (Tycho Brahe's astronomical assistants stayed with him for months at a time). Family members also supported such scientific work, as illustrated by Elisabeth Koopmann (1647–93), second wife of the eminent astronomer Johannes Hevelius (1611–87). She became his primary collaborator after their marriage in 1663 and she carried on their joint work after his death.

Trading companies were increasingly aware that specialized schools were necessary to give their employees technical training. Thus, in 1503, following the discovery of America, Queen Isabella I of Castile founded a hydrographic institution, the *Casa de la Contratación* (House of Trade) in Seville, the center for travel to the West Indies. Its responsibilities included the training of pilots, production of sea charts, and the design of nautical instruments. Instructors taught not only the use of these instruments but also elementary astronomy based on Sacrobosco's *Sphere*. Specialized schools also arose around this time in fields such as engineering or mining, where students learned practical mathematics as needed (cf. Chapter 6). Telling examples include the school of mathematics and engineering founded in 1600 in Leiden by Maurice, Prince of Orange, in the context of the war with Spain. The program designed by Simon Stevin (1548–1620) aimed at educating engineers in the Dutch language. The Royal Academy of Mathematics, established in 1584 by Juan de Herrera (1530–97) in Madrid, also stressed mathematics for civil and military engineers.

As mentioned above, from the sixteenth century onwards, new colleges arose outside the university setting with the explicit aim of educating wealthy urban elites. In the present context, the Jesuit colleges played a particularly significant role. As the first teaching order in the Catholic Church, the Jesuits made the founding, management, and staffing of colleges one of their highest priorities and most characteristic activities. Fifty years after the first college opened in Messina in 1548, the Jesuits, as a global intellectual elite, managed some 250 colleges and more than 500 worldwide by 1650. That number grew to more than 600 institutions of higher learning by 1700, with more than 200 in Central Europe alone. During the eighteenth century, the Jesuits operated in addition about twenty-five astronomical observatories. The impact of their teaching, especially in the realm of the mathematical sciences, can hardly be overestimated. Many prominent scholars and mathematicians of the early modern period, including Descartes, studied at a Jesuit college and profited from a thorough and up-to-date education.

Another important setting for mathematical practice in the early modern period were the craft workshops, which offered material and mental resources to their apprentices under the guidance of master craftsmen. A noteworthy fact concerning the education of a master craftsman is that this lasted longer than that of a master of arts. Artists, goldsmiths, metalworkers, stonecutters, architects, carpenters, etc. needed basic geometrical knowledge to shape ovals, bishops' crosses, ornaments, and other elements of their craft. They learned this in workshops through the pedagogical practice of copying from models in the form of books or artifacts. The "creative labor" (Pamela Smith) of the workshops could lead to innovations in geometry, such as the construction of hitherto unknown curves which were later explored in the realm of scholarly mathematics. The invention of perspective, or methods of foreshortening by means of geometrical projection, was one of the most outstanding achievements (cf. Chapter 5). This gave rise in the sixteenth century to an important written tradition by author practitioners, who claimed recognition as geometers. However, perhaps the most striking achievement of practical mathematicians was the construction of instruments as tools for problem-solving (Bennett 1986, 1991). Innovative practitioners built devices for drawing maps, finding a latitude, foreshortening an object in a perspectival representation, etc.

The innovative contribution of skilled artisans was especially fertile in so-called trading zones. These were "arenas in which artisans and other practitioners (trained as apprentices in workshops or in hands-on instruction at, say, construction sites) and learned men (trained in Latin at universities and other institutions) engaged in substantive communication and shared their respective expertise" (Long 2015: 842). Petrus Ramus (Pierre de la Ramée, 1515–72) gave significant testimony of this fertile encounter of knowhow and book-learning after visiting Nuremberg in 1570. Ramus admired the interactions he witnessed there in local craft workshops regularly visited by the learned.

Another famous trading zone was the Venetian arsenal (or armory), which offered space, material and personnel to shipwrights who experimented with a variety of ship designs. At some point, Galileo acted as consultant in this project. More generally, arsenals developed into places focused on very different problems, but especially those related to the production of cannons and gunpowder. Some carried out ballistic experiments or trained cannoneers; others designed fortifications or supervised their construction. This practical work was accompanied by a wealth of writings, e.g. on artillery, ballistics, fortification, shipbuilding, etc. Mines and mining had close links with armories, since most weapons construction required iron or bronze. There was an increasing, militarily relevant demand for metal, which had to be satisfied, and which entailed increased investments (governmental as well as private) in mining and metal processing techniques. This also gave rise to new technical challenges (ventilation, pumping out water, disposing of excavated material), which in turn led to experiments and theoretical reflections. Eventually, an entire new field of knowledge emerged: subterranean geometry (cf. Chapter 6 and Morel 2022).

The early career of Galileo nicely illustrates various types of settings that were typical for mathematical practitioners. Galileo was a university lecturer at Pisa and Padua for a large part of his career (1589–1610), before moving on to become the Medici's court philosopher and mathematician in Florence, where he profited from their generous patronage. While in Padua (1592–1610), he would often visit the workshop culture of the Arsenal in nearby Venice and the glassmakers on the island of Murano, acquiring hands-on expertise with lenses (essential when he built a telescope in 1609). To bolster his salary—mathematicians were usually paid at the low end of the university scale—he designed and built mathematical instruments (a military compass), published a booklet explaining how to operate it, *Le operazioni del compasso geometrico, et militare* (Operations of the Geometric and Military Compass, Padua 1606), and tutored private students in its use. These were often noblemen or military men, whom Galileo would board and teach in his own household. Having made a significant move upwards on the social ladder when he joined the Medici court in 1610, he became a member of the Accademia dei Lincei in 1611; he even secured the favor of the Jesuit mathematical elite and later that of the new pope Urban VIII. Thus, Galileo was at home in various milieus and settings, as were many other early modern mathematicians.

MATERIAL AND INTELLECTUAL RESOURCES

The mathematicians of this era worked with many of the same material devices and intellectual skills as used by other contemporaries who were engaged in the production and dissemination of knowledge. These involved writing, drawing, and printing, as well as reading, annotating, and copying, whereas other skills specific to mathematics led to major shifts in mathematical practice (see below). Language was a key aspect in the processes of learning, teaching, and transmitting mathematical knowledge. As we have already seen, recovering and translating mathematical texts from the Greek and Arabic traditions was fundamental for the development of mathematics in the early modern period. The most prominent example, often called Archimedean revival by historians of mathematics, came after the recovery of Archimedes's works, which had an impact that can hardly be overestimated. However, most of the early translations and commented editions were in Latin, while it became increasingly clear that the essence of these texts was also in high demand in vernacular. This posed quite a challenge because often translations of mathematical terms did not exist yet in, say, Dutch, English, French, or German, so these had to be established or invented. The Dutch mathematician Simon Stevin (1548–1620) is famous for his systematic approach to this problem and he even invented a new word for mathematics in Dutch, *wiskunde*.

Mathematics, considered as a textual enterprise, shared with other disciplines rhetorical devices taught in early modern liberal arts faculties as essential parts of the *trivium*. Stevin, for instance, wrote a treatise on rhetoric, *Dialectike ofte bewijsconst* (The Art of Proof, 1585) and applied its precepts to the structure of his *Arithmetics* (1585). In a pedagogical context, some mathematicians composed their books in dialogue form, where a master teaches his student. An interesting example is Robert Recorde's *The Castle of Knowledge* (1556), which is otherwise famous for an early reference to Copernicus. When the student considers the idea that the earth might move as absurd, the master rebukes him as being "to yonge to be a good iudge in so great a matter" (Recorde 1556: 165).

Mathematical texts of the period were often interspersed with diagrams and images that had close connections with the text, rather than being mere illustrations. Pictures could contribute to a better understanding of the content and might even play an argumentative role on their own. At times, they were also a means to put inventions or new ideas into circulation (cf. Chapter 7). In the sixteenth century, they may have drawn new readers within a broader literate public searching for mathematical knowledge. Some early modern cosmography or astronomy books included paper instruments (called volvelles), i.e. printed images with movable parts. Readers could cut these out to build a three-dimensional scientific object (Karr Schmidt 2018). The first such printed instrument was probably in Regiomontanus's 1474 *Ephemerides*, followed by others in works by Johann Stöffler (1452–1518) and Johannes Schöner (1477–1547) (cf. Chapter 2). Unlike Schöner, who addressed university-trained astronomers, Peter Apian (1495–1552) prepared paper instruments in his *Cosmographia* of 1524 for use by mathematical neophytes. Apian thus offered the latter the knowhow and tools for making their own measurements and calculations with the inexpensive portable instrument printed in the text (Gaida 2016: 281). Printed instruments, whether singular sheets or parts of a book, could be cut out and pasted on cardboard or wood. They were very common in the sixteenth and seventeenth centuries as an alternative to rather expensive brass instruments.

Annotated copies of early modern books found today in libraries often provide evidence of how books were used pointing to their readers' interests. One can also sometimes gain an impression of when and where they circulated and who these readers were by comparing the handwriting of the marginal notes, which might appear in different languages, from vernacular to ancient Greek. Archives and research libraries often hold manuscript notes pertaining to extracts from books. Reading, annotating, and copying were important practices in the process of producing knowledge, also in mathematics. Copying was one of the methods used for learning, at the universities as well as in (arts or crafts) workshops. When made by hand, slight changes could eventually alter the sense of a work and even lead to new knowledge, whereas the production of identical printed copies facilitated the circulation and to a certain extent the stabilization of knowledge.

In a time when copyright did not yet exist, severe contests sometimes arose over priority issues. During the sixteenth century, mathematicians typically regarded their methods and tools as private property, which they preferred not to share with others. One of the most famous priority disputes concerned methods of solution of the cubic equation, an issue hotly debated by Cardano and Tartaglia. Another example from the seventeenth century was the debate between Fermat and Descartes over methods for finding tangents to curves, a controversy fueled by letters exchanged with Mersenne. Recent studies have shown that such controversies played an important role in the creation of early modern knowledge (Bodenmann and Rey 2013). Taking full account of polemical writings leads to a completely different image of the Republic of Letters than the standard picture of a polite transnational community collaborating across institutional, political and religious borders. As far as mathematics was concerned, controversial issues sometimes arose in so-called mixed mathematics (cf. Chapter 5). Thus, Bernard de Fontenelle (1657–1767), secretary of the French Academy of Sciences, once commented about debates over the maneuvering of ships: "As these kinds of matters are not pure Geometry, but depend on a very delicate mixture of [. . .] Geometry & Physics [. . .] it was not surprising [. . .] that the Geometers could split. And indeed the latter split" (*Histoire de l'Académie royale des sciences*, 1714: 107). Of course, they also split very often when discussing matters of pure mathematics, such as the nature of infinitesimals. The seminal inventions of the differential calculus by Leibniz and the fluxional calculus by Newton gave rise to heated controversies, culminating in their infamous priority dispute. The latter prompted Newton to develop and publish his earlier results, which thereafter became far more widely available. Although conceptually different, fluxions and differentials solved essentially the same problems—providing methods for computing tangents and calculating areas—and could be considered identical since the results obtained using one method were translatable in the other. Nevertheless, this priority dispute strongly marked the development of mathematics in the early eighteenth century (cf. Chapter 3 in Volume 4).

Among the resources that significantly changed mathematical writing, and indeed mathematical thinking and practice, the most important and lasting was without any doubt the development of symbolism (cf. Chapter 3). At the beginning of our period, symbols were nearly absent from mathematical texts, except for the use of majuscules representing points in geometry. Propositions were formulated as ordinary sentences, such as "to find two numbers such that their sum and the difference of their squares are given numbers" (Diophantus quoted by Cajori [1928/9] 1993: 89). Individual writers, to be sure, often used symbols in their private manuscripts or letters, just not in printed texts. The Italian algebraists and German *cossists* (from *cosa* meaning the unknown) introduced certain symbols as abbreviations. Thus, Rafael Bombelli in his *Algebra* (1572) wrote *p.* for plus and *m.* for minus, while Simon Stevin represented the unknown x by a circle with *1* inside, x^2 by a circle with 2 inside, etc. Before them, the Welsh mathematician Robert Recorde (1510–58) introduced the equal sign =, the plus (+) and minus (−) signs in *The Whetstone of Witte* (1557). These today standard notations, however, were slow to catch on. Thus, equality was often expressed rhetorically by the Latin *aequantur* (sometimes abbreviated aeq.), or the French *esgale*, the German *gleich*, etc. It is significant that the sign = did not again appear in print until 1618. It gained some popularity in seventeenth-century England, where John Wallis and Isaac Newton used it, but had to wait until the beginning of the eighteenth century to be fully accepted (cf. Chapter 1; Cajori [1928/9] 1993). Not until the time of Viète and Descartes did mathematicians begin systematically to use vowels for unknowns and consonants for coefficients of an equation. These authors no longer used symbols merely as a device for simplifying mathematical writing. Descartes's *Geometry* (1637) clearly shows that it allowed to introduce aspects of algebraic thinking into geometry and to rely on algebraic criteria, such as the degree of equations, when it came to finding procedures to solve them. Relationships between the coefficients of an equation reflect the nature of its solutions, and this knowledge eventually led to new questions concerning the existence and nature of the solutions to a given algebraic equation.

FORMS OF MATHEMATIZATION

Already in the sixteenth century, mathematics began to permeate quite a few aspects of early modern culture, with geometrical motifs often appearing quite prominently (see an example from the early 17th century in Figure 0.1). The purported motto over the door to Plato's academy, "Let no one ignorant of geometry enter here," as well as the widespread idea of God as a geometer, testify to this. Tracts on dancing or fencing abounded with beautiful geometrized illustrations, whereas highly geometrized forms became common in architecture, including military fortifications. Even in gardening and landscape design, the geometric approach became omnipresent, the most prominent example being the gardens of Versailles.

FIGURE 0.1. *Académie de l'espée* (Sword Academy) by Girard Thibault from Antwerp, [Leiden], 1628, Tab. XII. Purchased with income from the Jacob S. Rogers Fund, The Metropolitan Museum of Art.

Early modern practitioners, whether from academia or workshops, were well aware of the many potential and real uses of the mathematical sciences. Highlighting such uses was a very common rhetorical strategy throughout the early modern period (cf. Chapter 7). From the gamut of authors who dealt with this topic, one might consider Martinus Hortensius (1605–39) a representative voice. As a convinced Copernican as well as an admirer of Galileo, Hortensius took up a professorship in 1634 at the *Athenaeum illustre*, i.e. *Illustrious School*, in Amsterdam. The *Athenaeum illustre* opened two years earlier with the goal of offering preparatory courses for future university students and it later became the nucleus that led to the founding of the University of Amsterdam. The school also prepared students for careers in trade or administration, in particular by teaching the mathematical sciences. In his inaugural speech on the dignity and the utility of the mathematical sciences, Hortensius enumerated ten branches, from astronomy to practical arithmetic, and explained the utility of each of them. Simultaneously he stressed that they were of immense use for the study of law, medicine, and theology, as well as philosophy. Hortensius was not stingy in his praise of the mathematical sciences:

> That Goddess [= mathematics], guide of the mind and actions, whom we ought to rely on and obey, whatever we have in mind, whatever we conceive in our minds; never does she fail to shed the gleam of her noble majesty through the palace of mathematics. [. . .] Where other sciences, being full of uncertainty and conjecture, can neither reach the truth by themselves, nor produce a remedy for the falsities they contain, the mathematical sciences, lacking nothing, suffice for themselves; content with the guidance of nature only, they hunt and capture truth itself." (Imhausen and Remmert 2006: 95)

Instances of mathematization abound in the early modern period, not only in the academic realm but also in works involving a more practical perspective. Natural philosophy—particularly when undertaken in mechanics, optics, physics, and statics—experienced an increasing influx and use of mathematical tools and methods, though not always on the cutting edge. As described in Chapter 5, Galileo, when analyzing the phenomenon of free fall, relied on standard, less suitable mathematical methods (the Euclidean theory of proportions). Such problems notwithstanding, processes of mathematization in fields like optics and statics increasingly characterized the study of nature. Many contemporaries took notice of this shift, as did Galileo, a fervent propagator of the use of mathematics for the study of nature. We find this theme nicely illustrated by the engraving on the title page of his *Il Saggiatore* (cf. Chapter 7). Pursuing this Galilean thread a bit further, one notes the use of mathematics in sixteenth century music theory. The lutenist and music theorist

Vincenzo Galilei (1520–91), Galileo's father, studied acoustics experimentally, supported by his young son. They studied, in particular, the musical properties of stretched strings and described these mathematically. Vincenzo Galilei may well have directed Galileo on a path of research that combined experimentation with a mathematical analysis of the results.

Giovanni Alfonso Borelli (1608–79) was a prolific scholar who taught at the universities of Messina and Pisa. As a staunch follower of Galileo's mathematization program, he published work on the fringes of mechanics and physics as well as in biology and medicine. His two-volume *De motu animalium* (On the Movement of Animals, Rome 1680) was a significant contribution to what we today call "biomechanics." Borelli conceived of humans and animals as muscle-skeleton-machines and argued that the (mathematical) laws of mechanics as proposed by Galileo not only held for inanimate matter but also for living beings. God was a geometer, as Borelli made crystal clear in the preface dedicated to Queen Christina of Sweden (1626–89):

> Similarly, the operations of animals are carried out using instruments and mechanical means such as scales, levers, pulleys, winding-drums, nails, spirals, etc. The scientific knowledge of these means is the province of geometry. Thus, it is true that, in building the organs of animals, God exerts geometry. To understand them we need geometry, which is the unique and appropriate science to enable one to read and understand the divine book written on animals. (Borelli 1989: 2)

Optics entered Galileo's field of interests as a practical mathematician when he built a telescope in 1609, the device that launched his fame. With it, he made the stunning astronomical observations published in his *Sidereus Nuncius* (Starry Messenger, Venice 1610). While he himself did not explore the theory behind his telescope, Kepler did. Soon after reading the *Sidereus Nuncius*, he worked out this theory and published it in his *Dioptrice* (Dioptrics, Augsburg 1611), thereby providing a mathematical framework for optical instruments (cf. Chapter 5).

Finally, as is well known, his findings eventually led to a conflict with the Catholic Church over the heliocentric system, especially since Galileo argued that in the study of nature mathematical truths should take precedence over theological ones. Less well known is the fact that among leading Jesuits, even though they rejected the movement of the earth, there was considerable common ground between exegesis and the mathematical sciences. Prominent Jesuit scholars, such as Christopher Clavius (1538–1612) and Athanasius Kircher (1602–80), used mathematical arguments to interpret specific passages of the Bible, and their Jesuit brothers usually accepted their conclusions. Clavius, for instance, calculated the maximum number of fixed stars that could cover the outer sphere of the universe and came up with the astonishing number of 71.209.600. Biblical exegetes quickly accepted this figure, notwithstanding the fact that the standard interpretation of the Bible suggested that the number was infinite.

Mathematics was present in many other fields of knowledge, impossible to enumerate here. In this volume, the potential uses of mathematics as a resource for various fields of inquiry and practical applications are described in many chapters, from the contexts of everyday numeracy (cf. Chapter 1) to its connections with technological change (cf. Chapter 6). Here, we briefly mention two related disciplines, geography and cartography. Both enjoyed a rich tradition stretching back to antiquity, familiar especially from Ptolemy's *Geography*. Indeed, this book on geography surely reached more readers than his famous *Almagest*, which required considerable mathematical sophistication. The *Geography* was a standard text in the Islamicate world, and it quickly made its way into the world of printed Latin texts from the fifteenth century onwards. Many scholars, such as Gemma Frisius (1508–55) and Peter Apian, studied it, while reworking the idea of longitude and latitude grids in mapmaking. Modern mathematical cartography began, however, with the work of Gerard Mercator (1512–94), remembered today in connection with the Mercator projection (cf. Chapter 5). In France, Oronce Fine, the first occupant of the mathematics chair created in 1531 at the newly founded *Collège royal* in Paris, combined astronomy with cartography, surveying, and instrumentation.

Such new mathematical tools and concepts did not necessarily find instant support, however, as illustrated by the example of logarithms as a calculating tool in astronomy. Mathematical tools and arguments had always stood at the core of astronomical practice, but when Johannes Kepler published his *Tabulae Rudolphinae* (Rudolfine Tables, Ulm 1627) based on the astronomical observations of Tycho Brahe, many contemporary astronomers resisted using them. Some were no doubt discouraged to see that Kepler based his tables on elliptical planetary orbits instead of the standard circular ones, but they also balked at his use of logarithms in order to facilitate computation. Thus, when the Dutch astronomer Philipp Lansbergen (1561–1632), a Copernican as well, published his less accurate tables in 1632 (*Tabulae motuum coelestium perpetuae*, Middelburg 1632), they surpassed Kepler's in popularity for a while because the mathematical tools employed were more familiar.

From the perspective of practical mathematics, innovation often took the form of improved instruments in fields such as mensuration, surveying, timekeeping, navigation, etc. A crucial problem for seamen was finding latitude and longitude at sea (cf. Chapter 5 and Volume 4). Instruments such as quadrants, cross-staffs, astrolabes, etc. were adapted for piloting ships at sea. Knowing the local time, especially at night, was crucial for finding a ship's longitude. Here again, instruments based on geometry, such as the nocturnal (cf. Figures 0.2 and 0.3), described by Petrus Apianus (1524) (cf. Chapter 2),

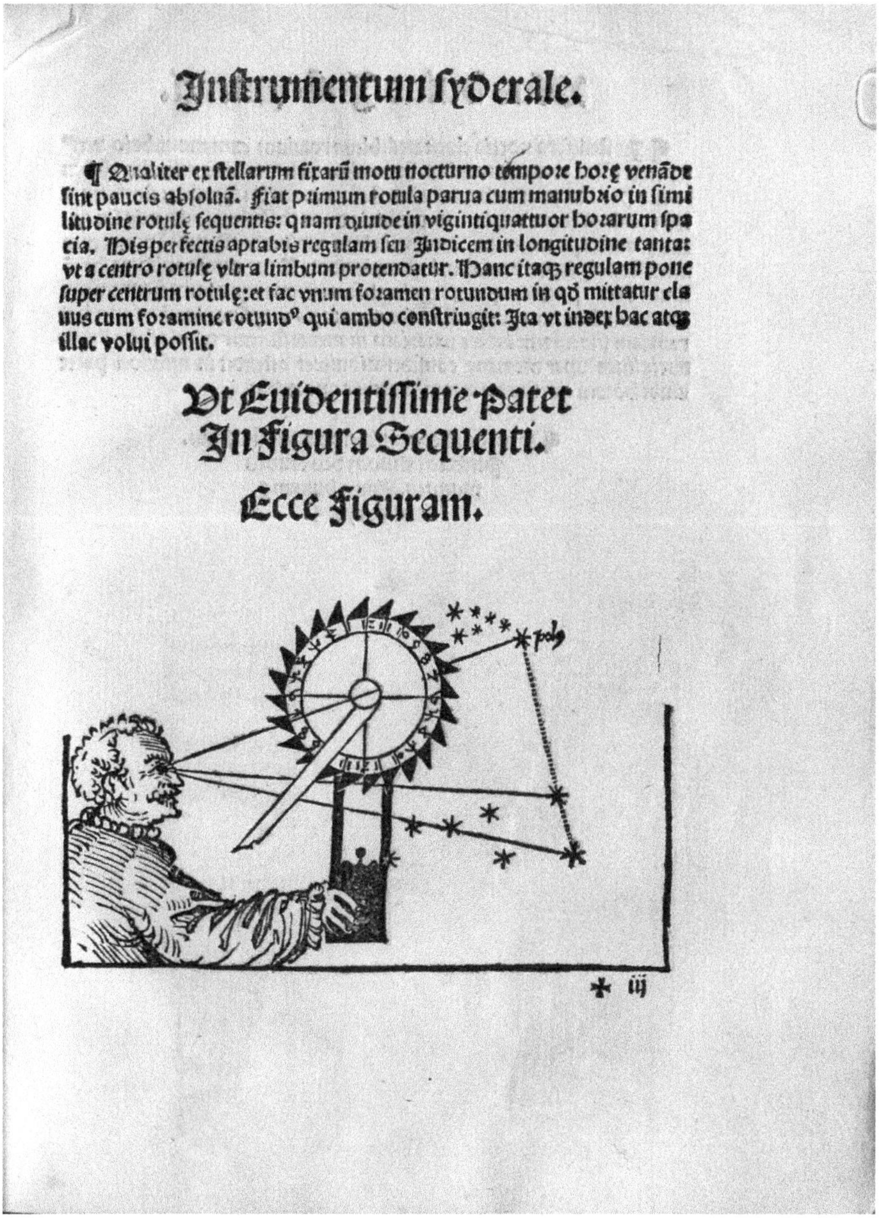

Instrumentum sydérale.

¶ Qualiter ex stellarum fixarū motu nocturno tempore horę venādę sint paucis absoluā. Fiat primum rotula parua cum manubrio in simi litudine rotulę sequentis: quam diuide in vigintiquattuor horarum spa cia. His perfectis aptabis regulam seu Judicem in longitudine tanta: vt a centro rotulę vltra limbum protendatur. Hanc itaq; regulam pone super centrum rotulę: et fac vnum foramen rotundum in qd mittatur cla uus cum foramine rotundus qui ambo constriugit: Ita vt index hac atq; illac volui possit.

Vt Euidentissime Patet
In Figura Sequenti.

Ecce Figuram.

✱ iij

FIGURE 0.2. The instrument "nocturnal" described by Petrus Apianus in his *Cosmographicus liber* (1524), second to last and last pages. Bayerische Staatsbibliothek München: Rar. 271.

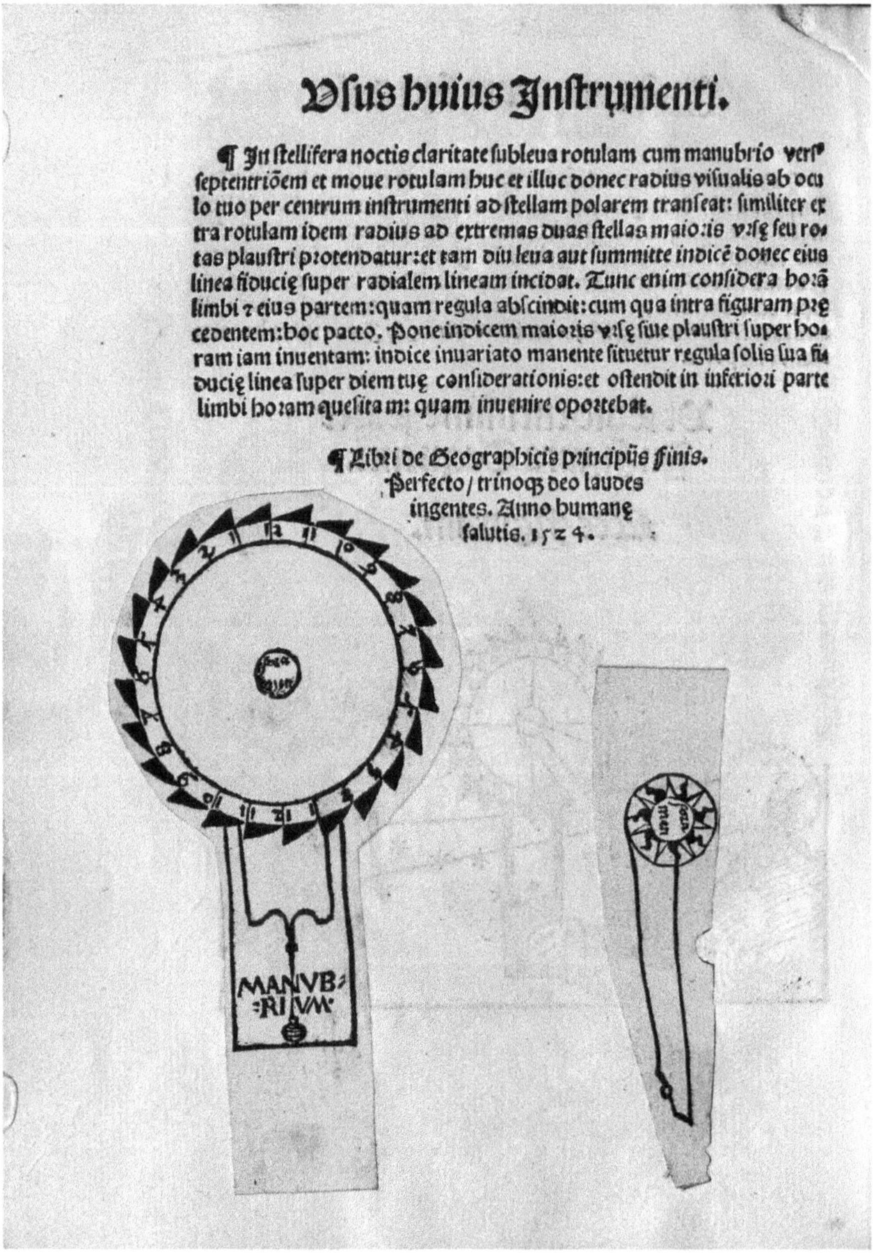
Vsus huius Instrumenti.

¶ In stellifera noctis claritate subleua rotulam cum manubrio vers° septentriōem et moue rotulam huc et illuc donec radius visualis ab oculo tuo per centrum instrumenti ad stellam polarem transeat: similiter extra rotulam idem radius ad extremas duas stellas maioris vrse seu rotas plaustri protendatur: et tam diu leua aut summitte indicē donec eius linea fiducię super radialem lineam incidat. Tunc enim considera horā limbi ꝛ eius partem: quam regula abscindit: cum qua intra figuram precedentem: hoc pacto. Pone indicem maioris vrsę siue plaustri super horam iam inuentam: indice inuariato manente situetur regula solis sua fiducię linea super diem tuę considerationis: et ostendit in inferiori parte limbi horam quesitam: quam inuenire oportebat.

¶ Libri de Geographicis principiis Finis. Perfecto / trinoqꝫ deo laudes ingentes. Anno humanę salutis. 1524.

FIGURE 0.3. The instrument "nocturnal" described by Petrus Apianus in his *Cosmographicus liber* (1524), second to last and last pages, with two printed pieces to be cut out. Bayerische Staatsbibliothek München: Rar. 271.

helped to determine the local time based on the observation of two or more stars at a given time of the year. Many of the same skills required for manufacturing instruments also applied in the production of accurate maps, which relied on knowledge of geometry (projections) and engraving methods. Just as cartography adopted mathematical principles, miners invented genuine geometrical methods to be able to map subterranean ground. Administrators started to compile bills of mortality and longevity tables needed for the computation of life annuities, an area of interest for social elites, which eventually contributed to the creation of an entire new branch of mathematics, namely statistics and probability theory (cf. Chapters 5 and 6).

The sixteenth century saw a flood of booklets addressing readers whose crafts relied on practical geometry. These writings typically opened with some basic results from Euclid, followed by a wealth of material dealing with constructions of special geometrical figures. The presentation was often in the style of recipes describing how to fashion materials such as wood, stone, or metal, whether in the realm of architecture or of artistic artisan crafts. They included questions on surveying, such as measuring the height of a tower located in an inaccessible terrain, about designing ornamental forms, such as for the capitals of say ionic columns. Sometimes they included or were exclusively devoted to various methods of perspectival constructions which were widely used in Renaissance art. Albrecht Dürer's *Underweysung der messung* (Instructions for Measuring, 1525) was an emblematic, though also highbrow instance of this type of manual. Although Dürer wrote it primarily as a training manual for craftsmen, his text also enjoyed some use at universities. For example, in the 1530s the astronomer Philipp Imser (1500–70), designer of a famous planetary clock (the Imser Clock, located today at the Vienna Museum of Science and Technology), used Dürer's *Underweysung* in his mathematical lectures at the University of Tübingen.

Geometrical knowledge also had many uses in military affairs during a time when warfare was an omnipresent reality in Europe. The art of fortification design changed considerably after the invention of gunpowder and the introduction of firearms. In Italy, theorists like Francesco di Giorgio Martini (1439–1502) and Baldassare Peruzzi (1481–1537) introduced fortresses with a polygonal structure and arrowhead bastions. This proved to be a popular model, widely studied by those officers who received training in the more technical aspects of modern warfare. Their education included geometry and arithmetic, as these subjects were indispensable prerequisites for artillery officers and military engineers. One of the latter was Luis Collado, who served with the Spanish army in Lombardy. In his *Pratica manuale di arteglieria* (Practical Manual of Artillery, Venice 1586), Collado explained that knowledge of arithmetic, geometry, and perspective was the very foundation on which artillery officers built their special understanding of the efficient use of cannons in warfare. The military was more and more regarded as an independent profession and saw the rise of an expert corps of professional officers with mathematical and technical skills acquired in special schools. During the Eighty Years War (1567–1659) between Philip II (1527–98), the Catholic king of Spain, and his Protestant possessions in the Netherlands—the longest armed struggle of the early modern period—Flanders became, in effect, Europe's academy of military science (Gonzalez de Leon 1996: 64).

Mathematics also played a crucial role in administration and trade during the early modern period. At the outbreak of the French Revolution in 1789, some 250,000 different units for measurements and weights were in use. This situation not only greatly hampered trade and central administration but also led to nearly insurmountable problems in assessing taxes. It also illustrates the need for simple methods for converting numbers given in two different systems of weights and measures, often by means of specific printed tables. In the financial sector, the international banking system relied heavily on mathematical methods, both old and new, as did a fast-growing insurance industry, which by the seventeenth century was issuing policies for overseas voyages as well as life insurance.

The sixteenth century also saw the rapid spread of a popular new print genre, namely almanacs. These circulated widely among the growing literate and numerate population (cf. Chapter 1), but especially at courts and among medical practitioners. Besides describing the movements of the stars and the planets, almanacs provided prognostications, noted weather patterns, indicated the dates for eclipses, and offered dietary and medical regimens. By the end of the seventeenth century, they had found their way into numerous households, where they served as calendars and notebooks for seasonal work, periodically occurring holidays, anniversaries, expenses, etc.

CHANGES IN EPISTEMOLOGICAL AND SOCIAL STATUS

As awareness of mathematics and the mathematical sciences expanded substantially from the fifteenth to the seventeenth centuries, their epistemological and social status also changed dramatically. Toward the beginning of the period, mathematics and the mathematical sciences occupied a low rank in the hierarchy of academic disciplines, especially compared with theology and philosophy, but also in relation to natural philosophy, and medicine. By the mid-seventeenth century, however, mathematization and mathematical modes of explanation had begun to dominate many fields of inquiry, especially those connecting with a direct impact on early modern societies. For many who practiced mathematics, this led to a new elevation in their social status. Mathematicians were no longer in the role of "Cinderella," having been transformed into the "Belle of the Ball" or even miracle workers. By around 1600, the mathematical sciences had become fashionable among social elites, whether in practice or just as table talk, whereas among Jesuits such expertise became a high distinction within the order. As the Nuremberg mathematician and expert on sundials, Andreas Schöner (1528–90), noted:

> there are now therefore many princes who even excel in the knowledge of the mathematical disciplines. Through their own effort they betake themselves to these studies as well as to a certain widening of the spirit. They delight themselves in these things. They design instruments and, indeed, I saw many of this sort designed by princes. They observe the motion of the heavens and perform the similar activities of their mathematicians. (Moran 1978: 189–90)

All in all, the social status of mathematicians rose to such an extent that the Jesuit Claude-François Milliet de Chales (1621–78), confidently asserted in the dedication to his *Cursus mathematicus* of 1674: "the other disciplines are plebeian, the mathematical sciences are regal."

This situation, however, contrasts sharply with what took place at the universities, where up to at least 1600 mathematics and the mathematical sciences remained subordinate to natural philosophy as well as the three core university disciplines: law, medicine, and theology. Mathematical methods had little or no significance within the framework of Renaissance Aristotelianisms, and consequently in natural philosophy, as then taught at the universities. When, during the latter half of the sixteenth century, a debate broke out in Italy over the certainty of the mathematical sciences (*Quaestio de certitudine mathematicarum*, cf. Chapter 4), the role and the status of mathematics became hotly discussed. An important outcome of this debate was that mathematicians throughout Europe increasingly took for granted that they produced and guaranteed the most certain knowledge attainable by humans. This development brought forth a new self-confidence among mathematicians as well as mathematical practitioners, an attitude that had echoes throughout the Scientific Revolution.

One thinks immediately of Galileo, who dared to challenge the claims of theology as being the only field to profess true and certain knowledge. For Galileo, mathematics not only represented just such knowledge, it made possible for humankind, by way of mathematical hypotheses, to truly describe and understand the natural world at large, the cosmos. As a follower of Copernicus, Galileo surely identified with the views expressed in the preface of *De Revolutionibus*, where Copernicus wrote: "mathematical things are written for mathematicians" (*mathemata mathematicis scribuntur*). Indeed, by Galileo's time, mathematicians throughout Europe agreed that to understand the natural world *without mathematics* was simply impossible. They spread that message and proudly advertised their new self-esteem while pursuing various strategies to legitimize the mathematical sciences and promote their status as scholars.

Their main vehicles for doing so were printed media, such as mathematical textbooks, guides to the practical use of mathematics, books of mathematical entertainments, editions of the classics, and encyclopedias (cf. Chapter 7).

The increasing epistemological and social status of the mathematical sciences proved to be a key prerequisite for the Scientific Revolution (Biagioli 1993; Dear 1995). It was fundamental to developments in the history of science and early modern European history as a whole, not least because it enabled the scientization of modern societies.

CHANGING QUESTIONS, NEW OBJECTS, TOOLS, AND METHODS

So far, we have described how the status of mathematics changed considerably from the time of the Renaissance. Along with literacy, numeracy progressed in European societies, as counting, measuring and weighing became standard components of elementary schooling, especially in urban environments. Mathematics was taught in ever more institutions and was part of the education of gentlemen and the *honnête homme*—a clear sign of its increasing social and cultural prestige, as legitimization efforts through print and speech proved successful (cf. Chapter 7). The era also witnessed a clear shift in the social organization of mathematics. Mathematicians were less dependent on wealthy patrons, namely at courts, to whom they dedicated their works. Instead, some gained positions in the academies founded during the second half of the seventeenth century, whereas many others held teaching positions. This shift in employment opportunities represented a step forward in the rather inchoate process toward professionalization. On the epistemological level, the potential power of mathematics to describe the natural world stood beyond question (cf. Chapter 4). Hence, the title of Newton's seminal opus, *Philosophiae naturalis principia mathematica* (1687), was scarcely debated and was widely accepted as self-evident. At the end of our period, nobody doubted that mathematics mattered.

In this final section, we focus on new mathematical objects and techniques, especially in the seventeenth century, when the world of creative mathematicians underwent dramatic transformations. We also consider the kinds of new questions they posed, as well as the nature of the objects they chose to investigate. Finally, what new methods did they invent in order to study them?

In the decades before and after 1500, many mathematicians were engaged in absorbing, studying, and editing ancient Greek mathematical texts. Two key figures in this regard were Federico Commandino (1506–75), a scholar of independent means from Urbino, and Francesco Maurolico (1494–75), a priest from Messina. Both approached classical Greek manuscripts critically, translating and correcting them, filling in gaps and thus providing new insights.

Commandino, for instance, gained one such insight while working on Ptolemy's *Planisphere* from an earlier Latin translation based on an Arabic version of the text. On studying it, he realized the connection between the stereographic projection of the celestial sphere and perspective studies made by architects in designing stage scenery (cf. Chapter 5). In another instance, when he was correcting an earlier printed Latin translation of Archimedes's *On Floating Bodies*, he came up with new ideas on centers of gravity. Maurolico, appointed in 1569 professor at the first Jesuit College in Messina, was by contract required to teach the theory of music as a branch of mathematics. He rearranged and translated parts of Euclid's *Elements* (published only in 1575), and in doing so modified Euclid's theory of proportions in Book V. These mathematicians combined their expertise as philologists with a strong desire to innovate, by going beyond the limits of classical knowledge.

It would be impossible to overestimate the influence of Euclid's axiomatic style as a model for mathematics written in the early modern period. A case in point is Niccolò Fontana (1500–57), known as Tartaglia, the Stammerer, who published the first translation of the *Elements* in Italian (1543). In a striking fashion, he employed the same Euclidean style in his *Nova scientia* (New Science, 1537), which dealt with gunnery and the trajectories of artillery fire. Tartaglia began with postulates, then deduced first propositions, and based all of his subsequent results on those he had already proven. In this way, he was able to develop a new science of ballistic methods, including firing tables. A broad Euclidean textbook tradition unfolded over this period, each author arranging the text in his own fashion, sometimes omitting the proofs. Heated debates took place throughout the seventeenth century as to whether algebra belonged in such textbooks. The name of Euclid was widely known among practitioners as well, many of whom referred to him as an unattainable horizon. Their approach favored an emphasis on geometrical instruments and constructions rather than on demonstrations.

In the mid-seventeenth century, the French mathematician Girard Desargues (1591–1661) set forth an alternative to Euclidean geometry; this later developed into what we now call "projective geometry." Like the technique of perspective invented by painters in fifteenth-century Italy, it dealt with projections and the changing shapes of a figure when viewed from different viewpoints. Think of a circular figure taking on elliptical shapes when viewed obliquely. In the sixteenth century, perspective was immensely popular among painters, architects, and stonecutters. These varied activities led to a familiarity with projections, since all dealt with objects and artifacts situated on ground planes and different movable elevations. Mathematicians, such as Commandino or Guidobaldo del Monte (1545–1607), also became interested in the mathematics related to these techniques. Desargues's work was rooted in praxis, and he wrote two treatises on stonecutting and sundials, respectively. He did not regard perspective as a simple procedure for foreshortening figures, however, but rather saw it as a method for transforming them, while maintaining certain relationships (the ones now called projective properties) (cf. Field 1997: 202). Desargues laid the foundations of projective geometry in an awkwardly written booklet issued in only fifty copies, the *Brouillon project* (Rough Draft of Attaining the Outcome of Intersecting a Cone with a Plane, 1639). Although addressed to artisans, the approach he took in his booklet was theoretical. He introduced the concept of points at infinity (cf. Chapter 3), which arose from his theoretical analysis of perspective. Even if some of his fellow mathematicians appreciated the *Brouillon project*—Pascal referred to it in his writings—it was unsuccessful and remained dormant until the nineteenth century, when Jean-Victor Poncelet (1788–1867) revived projective geometry.

One of the major developments of the early seventeenth century was the creation of symbolic algebra. During the Medieval period, Islamic mathematicians introduced an entirely new branch of mathematics, a "scientific art" they called "al-jabr," by which we recognize the name algebra. Its purpose was to solve problems that required finding an unknown quantity in numerical or geometrical problems. Greek mathematicians had earlier posed and solved such problems, as had their counterparts in Mesopotamia and India even earlier. Yet, it was a Persian polymath active in ninth-century Bagdhad, Muhammad ibn Musa al-Khwārizmī, who formulated the basic ideas of algebra. His *Book of Algebra* enjoyed a large circulation that extended to Western Europe. In it, al-Khwārizmī considered three kinds of quantities, namely simple numbers, the root (or thing, i.e. the unknown), and the square of the root. Using this terminology, he listed six basic types of (quadratic) equations and gave procedures for solving each of them. The algebraic art then consisted mainly in transforming equations involving one or more unknowns into one of these standard forms. This somewhat schematic description, however, does not render justice to the rich and deep insights and results of the Arabic tradition in algebra

These new mathematical methods arrived in Europe beginning in the early thirteenth century, following two main pathways: the first went along the trade routes terminating in southern Spain, the other passage led via Sicily to Italy. In the early sixteenth century, a group of Italian algebraists made significant advances in algebra, culminating with methods for solving third and fourth degree equations by radicals. Note, however, that Omar al-Khayyam (1048–1131) had already treated and classified cubic equations—written in rhetorical form, such as "Cube and sides equal a number"—in his *Algebra*, which he completed in Samarkand (Uzbekistan). He also solved a cubic equation by means of two intersecting conic curves, which produces the solution as a length, not a number.

What the Italians achieved was first published in 1545 in Gerolamo Cardano's (1501–76) *Ars magna* (Great Art), although Scipione del Ferro (1456–1526), professor at the University of Bologna, had given a solution earlier, but he kept his rule and method secret. Just before his death, he transmitted it to one of his pupils, who sought to turn it into hard cash by proposing the solution of cubic equations in public competitions with Tartaglia. The outcome was a hard and bitter quarrel among the protagonists. Cardano, although using geometrical terms in his text without any symbolic language, was able to express the solution in terms of the coefficients of the equation, today written symbolically as Cardano's formula. His approach was a major step towards a theory of equations. Yet, in one case, called "irreducible," he hit on complex numbers, as the formula included the square root of negative numbers. Although considered meaningless at this time, examples showed that the equation could nevertheless still have real roots. As a consequence, his compatriot, the hydraulic engineer Rafael Bombelli (*c.* 1526–73), introduced negative and complex numbers formally in his *Algebra* (1572), and he tried to calculate with them. This represents a first important extension of the number concept, which had hitherto included only positive real numbers.

A major shift in algebra occurred after François Viète (1540–1603), a French lawyer, created in his spare time what he called a new *analytical art* (In artem analyticem isagoge, 1591). Although his motivation was to recover the "analysis" of the Ancients, the main outcome of his efforts was the creation of a symbolism for algebra. This not only changed the style of algebraic presentation, it also had a deep impact on mathematics in general (see above and Chapter 3). The centuries-old practice of rhetorical algebra now gave way to a new symbolic algebra, that is algebra, as we know it today. Viète's main source, when creating symbolic algebra, was Pappus's *Mathematical Collection*, written in the fourth century and translated into Latin by Commandino in 1588. This collection was the main source for early Greek writings that would have otherwise been lost. It mentioned a hitherto unknown geometrical method called "analysis," which Viète wanted to recover as, in his eyes, with it one could solve any problem (*nullum problema non solvere*). He was largely unaware of parallel investigations already undertaken in Islamicate societies. Viète's analytical art, or symbolic algebra, achieved centuries after the creation of "al-jabr," in fact completely changed the face of mathematics.

René Descartes (1596–1650), in his influential *Geometry* (1637), applied Viète's analytical art to geometry (on Descartes, cf. Chapters 3 and 4). He translated algebraic operations into the language of geometry and, conversely, treated geometrical problems in an algebraic way. Thus, he represented geometrical objects, especially curves, by algebraic equations. This allowed him to extend the domain of geometry by systematically constructing curves with reference to their associated algebraic equations. To introduce new objects following Descartes's approach also meant to ask new questions, like, what devices are allowable for constructing new types of curves? Using new algebraic techniques in geometry also opened a foundational debate about what one means by a valid solution to a geometrical problem: a construction, as was the usual answer, or simply an equation? In general, Cartesian geometry enabled mathematicians to tackle old problems hitherto considered unsolvable (Bos 2001).

Some of the most popular curves in the seventeenth century were the familiar conic sections—circle, ellipse, parabola and hyperbola—already investigated by ancient geometers. Another rather simple curve, though, first gained attention in the early modern era. This was the cycloid, the path of a fixed point on a circle as it rolls along a straight line. While conic curves provided a good fit for describing the orbits of planets, they proved inadequate when it came to other natural phenomena. A simple example considered by Galileo was the so-called catenary curve, which arises from the question: what is the shape of a chain that hangs freely suspended at two of its points? Galileo understood that, while this curve seems to have a parabolic shape, in fact it is not a true parabola. It was only toward the end of the century, however, that Johann Bernoulli (1667–1748) showed that the true shape of a catenary curve could be described by a hyperbolic cosine.

In 1658, Blaise Pascal (1623–62) drew the attention of his mathematical community to the cycloid by posing a series of problems concerning this curve, such as finding its area (quadrature), center of gravity, etc. This prompted Christopher Wren (1632–1723), the architect of St. Paul's Cathedral who oversaw the rebuilding of London after the great fire of 1666, to announce that he had been able to rectify the cycloid, i.e. to find a straight line equal in length to the arc of the cycloid. This achievement was particularly significant in that it suggested to the mathematical community that solving such rectification problems was feasible. The Dutchman Christiaan Huygens (1629–95), one of the major figures of seventeenth-century mathematics when King Louis XIV appointed him to the newly founded Paris Academy of Sciences, used the cycloid in his design of a pendulum clock (published in 1658 and 1673, cf. Chapter 5). The key mathematical technique that emerged from this work was the generation of curves as evolutes and involutes (which turn out to be identical for cycloids). Originally defined by Huygens in mechanical terms, this afterward became a purely mathematical theory (cf. Chapters 3 and 5).

Wren's rectification of the cycloid raised the hopes of mathematicians that attempts to square curves—i.e. to compute the areas bounded by curves—might also be successful. The oldest and most famous quadrature problem was to construct a square equal to a given circle, or simply to square the circle, a problem that hinged on the mysterious ratio of its circumference to diameter represented by the number π. All through the seventeenth century, efforts to solve this and other quadrature problems only produced solutions for singular cases. Fermat systematically used symbolic algebra and heavily relied on Cartesian analytic methods to square the parabola. He and other mathematicians of the period, such as John Wallis (1616–1703), eventually succeeded in squaring higher order parabolas. Gilles Personne de Roberval (1602–75) was able to square an arch of the cycloid, using techniques inspired by a simplified method of exhaustion (cf. Introduction to Volume 1). The Italians Bonaventura Cavalieri (1598–1647) and Evangelista Torricelli (1608–47) introduced tiny indivisible quantities into geometry, imagining a figure as being composed of infinitely many such entities (cf. Chapter 3). Torricelli used this method of indivisibles to compute the (finite) volume of an infinitely long solid (Torricelli's Trumpet). Moreover, seventeenth-century mathematicians invented a large array of techniques allowing to determine tangents and normals to curves, to find minima or maxima of algebraic expressions, to compute volumes of solids, etc. An underlying question seemed to hang in the air: was there perhaps a general method that connected solutions of area problems to finding tangents to a related curve? Isaac Barrow (1630–77), Newton's professor at Cambridge, eventually realized that computing areas (or squaring curves) and finding tangents to curves were inverse operations, but he lacked the analytic methods of Fermat or Roberval to develop a general algorithm or calculus. This was to be introduced by Newton and independently by Leibniz.

The problems mentioned above, and more generally those involving curvature, required to be solved by techniques which went beyond Descartes, namely infinitesimal methods, which include the consideration of infinitely many, but also infinitely small quantities. Kepler's techniques for gauging wine casks involved such infinitesimal considerations. Although ancient Greek geometers had used infinitesimal methods as heuristic tools, these were banned in the formal texts familiar to early modern mathematicians. Since the Archimedean revival, following the first Latin translations of his work in the sixteenth century (1544, 1558), mathematicians resorted heavily to his methods in order to justify their arguments. They were often reluctant to use explicitly infinitesimal techniques, which many considered non-rigorous. These issues led to very heated foundational debates. Indeed, Archimedes had found ways to handle or better to obviate formally infinitesimal quantities, such as the method of exhaustion. Note that the idea of approaching a curvilinear figure by inscribed or subscribed polygons, which is at the core of this method, was considered to exhaust the figure only in the seventeenth century. A Flemish Jesuit, Gregory of Saint Vincent (1584–1667), in a work, *Opus geometricum*, 1647, devoted to squaring the circle used the word "exhaust" in this sense.

Infinite series, already present in Archimedean exhaustion methods, were to become a fundamental mathematical tool in seventeenth-century infinitesimal calculus. In his *Arithmetica infinitorum* (Arithmetic of Infinitesimals, 1656), John Wallis, Savilian Professor at Oxford, transformed the geometrical quadrature procedures of Cavalieri and Torricelli into purely arithmetical ones relying on interpolation techniques. While Cavalieri had been able to compute the area under the curve $y = ax^n$ for an integral exponent n, Wallis introduced fractional and negative exponents. This gave him a method for squaring the unit circle, represented by the square root of $1 - x^2$, which Wallis cleverly wrote as $(1-x^2)^{1/2}$.

Young Isaac Newton saw how to generalize Wallis's method. This led him to his first mathematical invention, the binomial theorem for fractional powers, i.e. the expansion into an infinite power series of the expression $(a + x)^n$, where n is a fraction. Newton then extrapolated his result to negative powers. As he was aware that the area under the curve $(1 + x)^{-1}$ is $\ln(1 + x)$, he could express that area as a power series, and by termwise squaring (or integrating) he could also express the logarithmic curve $\ln(1 + x)$ as a power series. This provided Newton with a powerful tool to represent a large number of curves—algebraic as well as transcendental like the logarithmic curve—by infinite series. Newton viewed the latter as infinite polynomials, whereas for Descartes the only allowable curves were those given by (finite) algebraic equations (cf. Bos 1980 [2000]; Jahnke [1999] 2003: chs 2–3).

Furthermore, Newton linked infinite series to what may be considered one of the most powerful inventions of the seventeenth century, namely the calculus. He understood that solutions to tangent problems (differential calculus) could be used to solve what he recognized to be the inverse problem of computing areas (integral calculus). In the 1660s, he developed his fluxional calculus (cf. Chapter 3) which gave an algorithmic procedure to determine the rates of change of changing magnitudes. Few, however, knew these methods in 1687, when he published his seminal *Philosophiae Naturalis Principia Mathematica*. Newton presented his masterpiece in classical geometrical language, following his philosophical agenda, which ranked ancient mathematics very highly. He thus avoided infinitesimal quantities, deeming these heuristic tools, and opted instead for a geometric limit procedure. He called this the method of first and last ratios, which he used to extend the principles of ancient geometry beyond simple curves to more general curved lines generated by continuous flow. Newton was able to solve a whole set of problems involving such motions—in planetary orbits, fluid flows, and in bodies acted on by central force laws, both in a void as well as in resisting media. All of these physical motions were subject to mathematical laws, which Newton elaborated in Books I and II of the *Principia*, and then applied to astronomical phenomena in Book III.

Leibniz was the first to publish the calculus, which appeared in a short paper in the October 1684 issue of the journal *Acta eruditorum* (cf. Chapter 3). Leibniz's approach differs significantly from Newton's. He is interested in infinitesimal differences between successive values of a variable, written dx, which fit the framework of symbolic algebra. In fact, these new symbols obey many of the same rules as ordinary variables. For example, if $y = 3x$, then these two variable quantities have a fixed ratio $y{:}x = 3{:}1$, as do their differentials, $dy{:}dx = 3{:}1$. Leibniz introduced the standard algorithm for the calculation of the differentials, and he put the advantages of his "new method" to the fore in applying it to several well-known examples. He thus showed how to easily obtain results that were derived earlier using more cumbersome and less general methods. Leibniz's paper was difficult to read, not only because of its innovative character, but also because of numerous misspellings due to the printer's lack of familiarity with the new mathematical symbolism. It took several years before Jacob and Johann Bernoulli, the foremost followers of Leibniz in Europe, were able to demonstrate its potential by solving new, ever more difficult problems. Leibniz began his research by exploring the properties of numerical series of differences. This exploration implied an idea which was to play a central role in Leibnizian calculus: the sum of an infinite number of terms b_n can be achieved via the difference sequence $a_n = b_{n+1} - b_n$. Integral calculus was then based on techniques for adding infinite sums of infinitesimally small quantities (or differences) to obtain a finite result.

The development of symbolic algebra, its application to geometrical objects like curves, and the invention of infinitesimal calculus were profound shifts inaugurated by several leading mathematicians over the course of the seventeenth century. They deeply transformed the presentation of mathematics by means of formulae, considerably expanded the realm of problems amenable to the new methods, and deeply anchored the trust in mathematics as an efficient tool to study physical phenomena. In the eighteenth century, the interaction between mathematics and mechanics intensified greatly. Continental mathematicians, beginning with Jacob and Johann Bernoulli, together with their student, Leonhard Euler, elaborated the mathematical methods of the Leibnizian differential and integral calculus. This new mathematical language soon became the modern vernacular for solving problems in mechanics.

We close this survey of early modern developments and contexts in mathematics by pointing to the fact that many mathematicians of this period were well aware of the various traditions they drew from, above all the works from the Greek and Islamicate traditions. Often they explicitly referred to these traditions, occasionally with great veneration for specific mathematicians (as in the case of Archimedes). Thus, it is not surprising that the history of mathematics and the mathematical sciences was cultivated as a specific interest throughout the period (cf. Chapter 7). Regiomontanus began his lectures in Padua in 1464 with a survey of the history of the mathematical sciences (Swerdlow 1993). The Italian humanist mathematician Bernardino Baldi (1553–1617) wrote biographies of more than 200 mathematicians, *Le Vite de' matematici*, ranging from Greece over the Islamicate world to Europe in his own time, with Archimedes receiving by far the most extensive treatment. While some have called Baldi the first European historian of mathematics, many mathematicians and scholars in the early modern period wrote about its history (Rose 1975; Goulding 2010), testifying to the fact that mathematicians were proud of their heritage and traditions and keen on cultivating them.

CHAPTER ONE

Everyday Numeracy

Everyday Numeracy through the Mirror of Textbooks

MARYVONNE SPIESSER

In Shakespeare's *The Winter's Tale*, Clown, the good-natured son of an old shepherd, though not the brightest of fellows, tries to calculate how much he could earn from selling his sheep fleece: every eleven sheep yield one *tod* (28 pounds of weight) of fleece; each *tod* brings in a pound and a shilling. Now, how much would he get for fifteen hundred fleeces? To this question, Clown answers: "I cannot do it without counters." We could have quoted many excerpts from literature that illustrate what the expression "everyday numeracy" refers to, namely the ability to use numbers to solve problems in all aspects of life, whether for personal or household needs such as home accounting, for recreational interest in numbers or for professional needs. All protagonists in this chapter are users of elementary arithmetic, albeit to vastly differing extents and purposes.

In early modern Europe, during a period of great technological changes, those "who know by doing and those who know by thinking" (Cormack 2017: 2) were not in opposition; more and more people were using mathematics for technical or practical purposes (trade, surveying, navigation, etc.), many educated men were widely interested both in theory and practice. The study of numeracy therefore takes place in this context of the rise of practical mathematics.

Number representations are at the core of numeracy. During the so-called European Renaissance, Hindu-Arabic calculation—already well established in the Islamic world—was spreading unevenly throughout Europe, but had yet to supplant its competitors. The ongoing evolution of the various methods of calculation was crucial: this is why we will focus on numeracy within the Western scientific tradition. Even within this limited framework, part of the difficulty lies in the heterogeneity of people's numerical skills: the local shopkeeper does not have the same needs as the international merchant, accounting techniques are more sophisticated in large Italian cities than in most other European areas, the literacy rate is higher among the urban population, etc.

The first two sections tackle the following questions: What were the techniques and devices in use to solve problems involving numbers? How could people become numerate? The third section addresses the issue of numeracy in daily (professional or private) life.

Obtaining a clear picture of the daily practice of arithmetic is not easy. Scattered testimonies, documents such as ledgers of account or reckoning devices throw some light on the forms and the unfolding of numeracy. But the most accessible and stable information has been passed down to us through "textbooks," a term covering different kinds of documents, which increased in number and diversified over time. These include teachers' or students' handbooks, manuscripts, or printed books, drafts, etc. It is chiefly through the analysis of these texts that we will address the issue of everyday numeracy.

COMPETITION BETWEEN RECKONING SYSTEMS

How can a person memorize numbers and perform calculations, whether he/she is literate or not? Unless one can make a mental calculation, which limits the possibilities and difficulties of the task, the natural alternative is to aid one's memory with an instrument (in a broad sense). Various inventive solutions have been implemented throughout history, including instruments that allow one to solve problems without needing to directly engage with mathematics.

Today, the Hindu-Arabic numerals, which were transmitted to Europe via Islamic science in the Middle Ages, have spread all over the world. But, though this written system for reckoning eventually became universal, it did not displace other older ways of doing arithmetic. Throughout the early modern period and even later, Hindu-Arabic numerals were used alongside the older method of calculation on the "abacus" (reckoning with counters). We discuss below the competition between these two different systems and their respective strengths and weaknesses, taking into account geographical areas and chronological evolutions. Moreover, we must not overlook a third counting system, still much in use during the early modern period: reckoning with fingers.

The first "instrument" employed in counting is accessible to all: the human body. Many parts of it have been used since people first began reckoning. All over the world, the use of fingers was common practice from ancient times—

for instance, in transactions—which still persists today, helping people to keep a number in mind while performing a calculation. This type of method entails both a visual representation of numbers on the hands and the use of memory for mental arithmetic: finger-counting makes numbers visible but evanescent.

Finger-counting was described in many books from the sixteenth century, such as *The Ground of Artes*, published in 1543 by Robert Recorde (*c.* 1510–58). Recorde had studied mathematics and medicine, and he became one of the most influential authors of mathematical textbooks in English at that time. The book ends with “the art of nombrynge on the hand.” According to Recorde, this was a pleasant method both for its strangeness and secrecy as well as for its convenience. While this manner of reckoning had been used for over 2,000 years, it had never been taught in English. Although Recorde promised to explain how to perform the four operations on the hands, he failed to do so. Was it possible to keep track of long accounts without any other help than one’s fingers? It is rather easy to add or subtract numbers, but carrying out a multiplication is far more complicated and burdensome. Representing numerals and making computations are two different things and, for early modern writers, finger-counting looked rather like the memory of an ancient practice; it seems that, in Europe, one used it chiefly to relieve human memory.

During the early modern period, an abacus was typically a ruled flat surface (sometimes a cloth) marked with parallel lines placed in front of the user. From bottom to top, in a decimal system the lines represent the ranks of the units, tens, hundreds, etc. Identical counters could be placed on a line or between two of them. The number of tokens gives the number of units of the corresponding power of ten. A counter placed between two lines is worth five counters from the one below it (cf. Diagram 1.1). Thus, the abacus was a very suitable tool when working with Roman numerals, an additive system in which written calculation was almost impossible. Operations were carried out by moving tokens and replacing any group of ten on a line by one token on the line above it. This is why “counter-casters” (who effectively cast tokens on their table) could find the Roman notation superior to the Arabic. Books entitled *Arithmetica linealis* in Latin or *Rechnung auf den Linien* in German, which means arithmetic or computing on the lines, taught this method.

A particular case, commonly found in Western European textbooks during the sixteenth century, was the abacus with virtual lines. This used special tokens arranged regularly, one below the other, to determine the starting point of imaginary lines; these formed “the numbering tree” (in Figure 1.1 this is the column with the named counters). These devices were used to reckon with abstract numbers and for calculations involving money as well. To accommodate local currencies or measures that utilized a non-decimal system, there were specific values for counters representing half the upper unit (for instance, half a shilling for 6 pence, half a score of pounds for 10 pounds). Recorde explains in *The Ground*

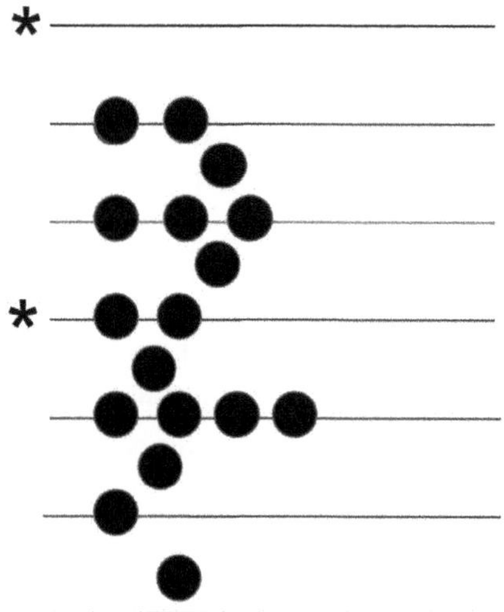

DIAGRAM 1.1. How to represent 287,965 “on the lines.” From Recorde 1543, *The Ground of Artes*: 118. Courtesy of Maryvonne Spiesser.

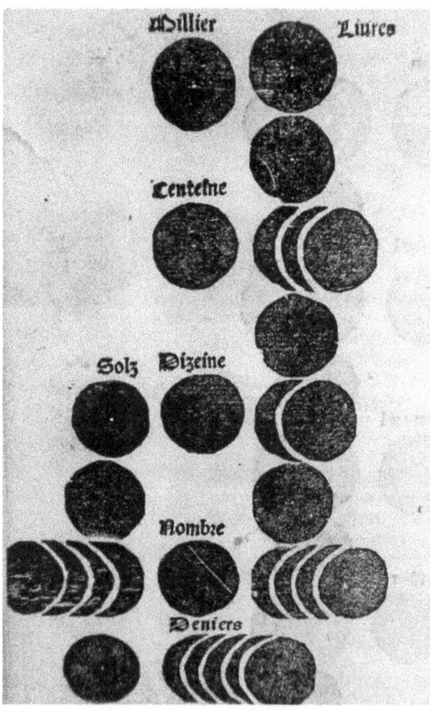

FIGURE 1.1. An addition of money (livres, sous, deniers) carried out with counters (Estienne de la Roche 1538, *Arismetique*, n.p.). Toulouse, Bibliothèque d’étude et du patrimoine. Photo by Maryvonne Spiesser.

of Artes that commercial and financial accounts did not follow the same disposition; the first were arranged vertically, with pence counters at the bottom, the second horizontally with pence on the right side. Figure 1.1 shows the arrangement of counters used to answer a question involving a merchant who receives 4£ (*livre*) 1s (*sol*) 2d (*denier*), 5£ 3s 1d, 70£ 5s 2d, 300£, 1500£ 10s 6d, and who wants to know the total amount. *Livres* are set on the right of the "tree," *sols* on the left and *deniers* at the bottom. Hence, 12d are worth 1s and 20s are worth 1£, the jetton at the bottom left is half a *sol*, that is 6d. The total gain is 1879£ 19s 11d.

A second familiar type of abacus consists of a table or frame with parallel rods fixed on it, along which beads can slide. Although this device lies outside our geographical scope, it is worth comparing it to the previous family because of a special feature: the number of beads on a rod is fixed. These instruments, which can be found in many places in the world (in Russia for example), were especially popular in China. The Chinese abacus or *suanpan* was not commonly used until the second half of the sixteenth century. The frame is separated in two parts (cf. Figure 1.2). There are five beads on each rod in the bottom deck, each carrying a value of 1, and two on each rod in the upper deck with a value of 5. The rightmost column represents units; to the left of it are the tens, hundreds, etc. This abacus was the basic tool used by merchants, who could perform their calculations with amazing speed. The *suanpan* soon spread

FIGURE 1.2. A *suanpan*. Photo by Maryvonne Spiesser.

through all of Southeast Asia and by the end of the sixteenth century, Japanese calculators adopted it. They called it *soroban*, bringing some changes, such as in the number of the beads in the same column: 4 on the lower part and 1 on the upper, which was sufficient for calculations in a decimal system. During the Edo period, which began in 1603, the *soroban* became popular with amateur mathematicians and enjoyed enduring success.

Arabic scholars played a crucial role in transmitting the decimal place-value system with its ten Indian numerals to Europe around the twelfth century (cf. Chapter 1 in Volume 2). The original glyphs underwent successive transformations before arriving in Europe, after which they continued to evolve. The main difference between this way of representing a number and an additive system like Roman numerals is that the value of a figure depends on its position. For instance, V *always* means five units, whereas 5 means five units in the number 25 but five hundred in the number 2567. When counting, say in tens, one repeats the procedure for units, thus the number 90 (or 9 tens) leads to the next higher number place: one hundred, which has no tens and no units. All this led to the invention of the figure 0, which bore different names during the Middle Ages like *cifra*, or the more colorful French *non riens*.

During the twelfth century, the Latin translations and adaptations of al-Khwārizmī's Arabic text on "Indian calculation" (ninth century) gave rise to a new kind of Academic Latin treatise called *algorismus* (from the Latinization of the author's name). The Hindu-Arabic system spread to Southern Europe, thanks to *algorisms* written in Latin during the thirteenth century and later in vernacular languages. First conducted on a board covered with sand or dust, but with intermediate calculations being erased, it progressively moved on to wax tablets (as late as the sixteenth century), then slate and finally paper, when this medium became cheaper and more common. In contrast to the other systems, one could record the steps in the calculation process on paper. The great Italian cities that dominated the Mediterranean trade during the Middle Ages were at the forefront of this new method of calculation. Then the decimal place-value system disseminated from south to north, as many methods for carrying out the four main operations emerged (cf. Figure 1.3).

If the efficiency of this place-value system seems obvious in retrospect, this process of dissemination did not occur quickly. Some maintained, for instance, that Italian merchants feared that their accounts would be more easily falsified. While Greeks and Romans of the classical world calculated with an abacus, there is no archaeological or textual evidence for a continuous use of it in medieval Europe. The counter abacus (with horizontal lines) did reappear, according to experts some time between the thirteenth and the fifteenth century, and almost exclusively in Northern Europe. This is something of a puzzle: did the counting board completely disappear from the Mediterranean only to reappear a very long time later on the fringes of the old Roman Empire? Other

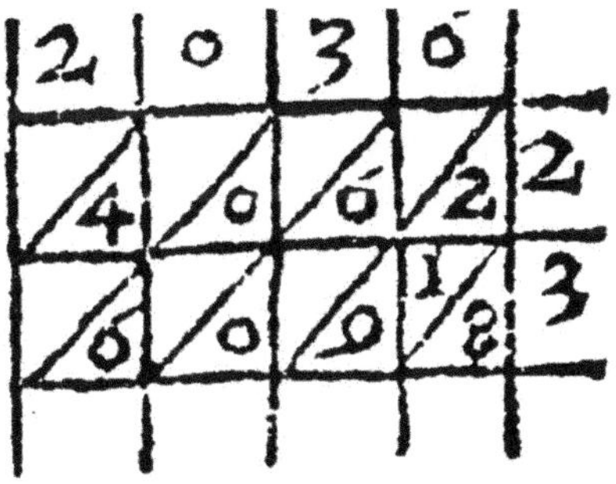

FIGURE 1.3. A multiplication by gelosia, or "To multiply by a checker table." Photo by Maryvonne Spiesser.

specialists do not share this point of view; some of them favor the view that the abacus was in continuous use in Northern Europe. Anyway, the abundance of material and written documents proves that counter-casting was widespread during the early modern period and even later, in countries such as England, Holland, Germany, Denmark, or Northern France. A large number of metal counters belong to museums and private collections, especially counters that originated from Germany where *Rechnung auf Linien* was highly promoted during the fifteenth and sixteenth centuries. One of the best known German reckonmasters, Adam Riese (1492–1559) claimed, based on his teaching experience, that young people who reckoned "on the lines" calculated faster than the proponents of written calculation. The first computation book he wrote, in 1518, was devoted to the counting board as it is usually taught in all schools of computation. The second one, dated 1522, was entitled *Rechnung auff der Linihen und Federn* (*Computations on the Lines and with the Quill*). Recorde, by contrast, first taught written calculation and used it to explain numeration and operations with counters to his pupils. Among the artifacts that have survived from this era, one finds besides jettons, engravings and paintings which depict merchants working at their counting tables, whereas museums have preserved various devices, such as engraved tables. These look like ordinary ancient wooden tables, except that their tops have inlaid engraved patterns or a marquetry that transform them into calculating tables. Other devices such as counting mats have also been preserved. All these objects come from Northern European countries. Some of them were used for private purposes and may have stood in the homes of merchants, others were used to keep public accounts, still others belonged to a parish or to the chaplain of a cathedral. This stands in sharp contrast to the situation in Italy, where one finds no trace of any such devices. In fact, a Venetian patrician who died in 1495 asserted that during his lifetime counters were never used. This is not surprising, since Italy pioneered written calculation as early as the thirteenth century.

Arithmetic textbooks give a good indication of the reckoning methods that were then in use. While in Southern Europe textbooks from the fifteenth century were almost exclusively *algorisms*, those published in Northern Europe a century later still offered both types of reckoning techniques. In fact, these methods did not require the same skills. Illiterate people could learn by heart how to perform an operation by moving counters on a table but, as they could not write, they were unable to use Hindu-Arabic figures. Moreover, the abstract idea of a figure signifying "nothing" was certainly difficult to accept, while on the abacus "nothing" simply meant the lack of tokens. In the preface of *Le livre de chifres et de gects* (The Book of Figures and Tokens), an anonymous French textbook published in Lyons in the early sixteenth century, the author explains in the introduction that his book contains "first the arithmetic which is implemented with the tokens [. . .] for those who cannot write, then for those who can write is explained the arithmetic called by the figures." The same argument is given in the first printed Arithmetic in England (1537), *An Introduction for to Lerne to Recken with the Pen, or with the Counters.* Meanwhile, Recorde, when introducing "accounting by counters," asserts that it also serves literate people who "have not at some times their pen or tables ready with them"—not to overlook that paper was expensive and rare for a long time.

The famous engraving *Arithmetica* (Figure 1.4) appears in the book *Margarita philosophica* (Pearl of Wisdom), written by the German humanist Gregor Reisch (*c.* 1467–1525) and first published in 1503. It highlights the competition between counters and pen: *Lady Arithmetica*, her gown bedecked with Hindu-Arabic numerals, turns her smiling face towards Boethius, who represents written calculation while ignoring Pythagoras, who is toiling at a counting board with counters, thus portrayed as an old-fashioned process.

Yet, the movement towards written calculation was neither linear nor homogeneous, even if it ended up winning the game. During the sixteenth century, much bookkeeping continued to be done with Roman numerals, even when the final records were reported with Arabic numerals. The Lord High Treasurer of Elizabeth I of England transcribed economic documents from Hindu-Arabic back to Roman numerals for his own convenience. Recording results with Arabic figures was one thing, calculating with them is quite another. In an English administrative document dating 1600, amounts of money are reported in Roman figures, though the total at the bottom right is given in Arabic numerals. Yet, on the bottom left, one finds dots representing counters as they lay on the Table. Evidently, the calculations were performed on a board from data in Roman numerals, and the result was converted into Hindu-Arabic numerals for the record of accounts. In other documents, Roman numerals were used instead for official records.

FIGURE 1.4. Arithmetica, in Georg Reisch. 1508. *Margarita philosophica cum additionibus novis*. Basel: Schott. Bayerische Staatsbibliothek München: Res/4 Ph.u. 117.

The debates between the proponents of either of these two systems reflect contested narratives with respect to their purported efficiency. Generally, the practice of casting counters was favored when performing an addition with many numbers. As the French Lyons mathematician Jean Trenchant (sixteenth century) wrote, the adding of a large number of terms by pen easily led to errors because one had to memorize the numbers to be added beforehand. On the other hand, this method of reckoning allows one to keep track of intermediate results, which cannot be done by moving counters on the board. Written calculations also produce permanent records, which allow one to detect errors more easily. Another valuable advantage of the Hindu-Arabic system was its obvious superiority as soon as one had to perform multiplications and, even more so, divisions. Robert Hartwell, in the appendix of a seventeenth-century edition of the *Ground of Artes*, mentioned as an example of ignorant people those who “can but cast with counters.” And Jean Abraham, in his *Arithmetique*

(1616), responded to those who claimed that businessmen do not trust Arabic figures by saying that they would rapidly change their minds if they learned how to use them.

Nevertheless, the shift from a popular system to a technically superior one was not simply a matter of recognizing its superior effectiveness. It also depended on how many people used it and for what purposes, as noted by Stephen Chrisomalis:

> Once the Western numerals had achieved a critical mass of popularity among the newly emboldened European Middle class, it became likely that others operating within the same economic and communication networks would adopt the system. [. . .] The 'cost' of not using the popular system is greater than the advantage of using the technically superior one. (Chrisomalis [2009] 2011: 512)

Let us add that it is not easy to get rid of deep-seated sociological beliefs and practices, which also played a major role in this story. For instance, the perceived social and intellectual inferiority of women contributed to the idea that counter-reckoning was more appropriate for them than written calculation.

Levels of mathematical ability were quite heterogeneous among practitioners who had to use numbers. Multiplication and division were regarded as laborious chores, time-consuming, and prone to errors. To obviate such difficulties and relieve memory as well, multiplication tables were included in many practical arithmetic textbooks. By the seventeenth century, with the same purpose, printed books of tables began to appear, later on called *Ready Reckoners*. The latter included pre-calculated results of all kinds useful in professional life, as will be described briefly in the last section of this chapter.

Beyond printed aids, a variety of artifacts have been devised in different cultures for recording numerical results but also to perform arithmetical operations, as with abacuses. Knotted strings (like the *quipus* used by the Incas) or *tallies* were well-spread cross-cultural artifacts, created long ago to fulfill recording tasks like accounting. Tallies were originally pieces of wood on which notches were cut; the British Exchequer used this primitive means of account-keeping up until 1826. When split longitudinally this provided a record for both depositor and the exchequer.

John Napier or Neper (1550–1617), the Scottish mathematician, physicist, and astronomer, best known as the creator of logarithms, invented an instrument that reduced multiplication to addition: the "Napier's rods" or "Napier's bones." He described it in a work called *Rabdologia* (collection of rods in late Greek), printed in 1617. Napier's rods indeed enjoyed considerable success, all the way to China and Japan, an enduring popularity throughout the eighteenth century, and a revival in the late nineteenth century. It consists of a system of rods arranged to represent the *gelosia* method of multiplication (Figure 1.5).

Let us consider the simple example of 456 × 38 (cf. Diagram 1.2). Each rod contains the multiplication table for a digit from 0 to 9. We choose the rods 4, 5, 6 and arrange them as shown in the diagram with the sequence of digits 1 to 9 on the left side. The product of 456 by 3 (tens) can be read on row 3, adding the digits in each rhombus and starting from the right: 8, 5 + 1 = 6, 2 + 1 = 3, 1. The result is 1,368 (tens). Then the same process allows us to find the units by multiplying 456 by 8. If the sum of one rhombus is more than 10, you must carry 1 to the next rhombus. Adding the results for the number of tens and units yields the result 17,328.

Of course, the rods were chiefly of interest for large numbers, and Napier explained how to multiply 8,795,036,412 by 3,586,290,741.

Along with Napier's bones, other devices like slide rules were created that made use of more recent mathematics like logarithms. Blaise Pascal's (1623–62) invention of his calculator, opened the way for the construction of other mechanical calculating machines during the same century. Sir Samuel Morland (1625–95), inventor and natural philosopher, constructed a machine that was advertised in 1674 as suitable especially for "Ladies, that desire to look into their disbursements, or layings out, and yet had no time to practise in numbers" (Wardhaugh 2012a: 111).

Writing, in the ordinary meaning of the word, is only one of the instruments of knowledge. *Tallies*, for example, based on a principle analogous to a

FIGURE 1.5. Napier's rods. Photo by Patrick Guenette / Alamy Stock Photo.

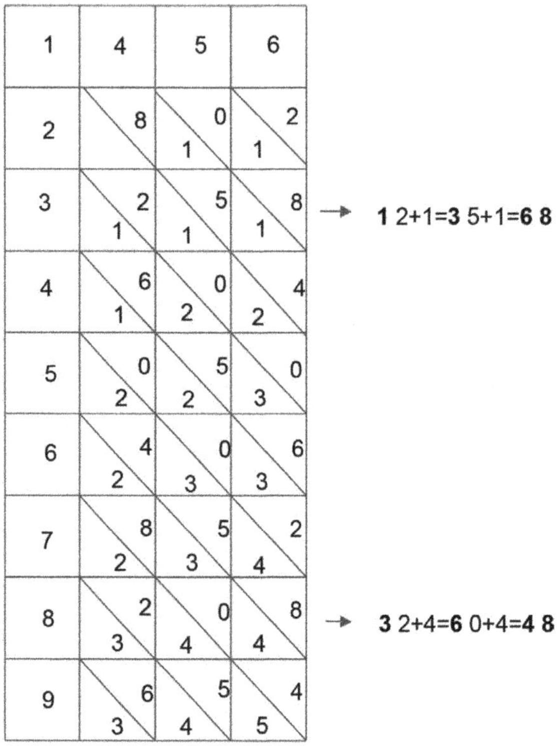

DIAGRAM 1.2. A multiplication performed with Napier's rods: 456 × 38. Photo by Maryvonne Spiesser.

rudimentary writing system, coexisted with other accounting methods and lasted well beyond the seventeenth century. As the long-drawn-out struggle between the abacus and the pen testifies, these practices evolved in response to a variety of factors and contexts.

ACQUIRING NUMERICAL SKILLS

The transformations of society that took place over the sixteenth and seventeenth centuries shook up the intellectual and educational landscape, as Humanism tended to dethrone medieval traditions and organizations. Educational structures underwent major upheavals, too, with the rise of secondary schools or colleges devoted to the study of the arts, which served as the basis of education for new elites. Some colleges promoted the Protestant Reformation, others the Catholic Counter-Reformation (among them the Jesuit colleges) in an atmosphere that turned the landscape of learning upside down.

Many thought of mathematics as dehumanizing and that it made a person inept in social life (Thomas 1987: 109). Its study was generally regarded as a low priority, despite local or individual exceptions. In England, for instance, the natural philosopher John Dee (1527–1608?), who was largely a self-taught mathematician, became an influential promoter of mathematical learning. In the Netherlands, Gemma Frisius (1508–55), a master of arts, doctor in medicine, and the most influential Dutch mathematician of the first sixteenth century, introduced practical mathematics into the curriculum. Along with his French contemporary, Oronce Fine (1494–1555), Frisius founded modern cartography and geodesy. In 1540, he published a Latin textbook (*Arithmeticae practicae methodus facilis*), which became one of the outstanding books for Latin schools in the Netherlands, where almost no mathematics was taught and surely no practical mathematics had been taught earlier. The idea that studying mathematics not only benefits the mind, but also is useful for everyday life owes much to these pioneering scholars.

In this changing environment, one faces many obstacles in trying to form a picture of training in elementary practical arithmetic. Source material is sparse and there are great variations between places and countries over the period, not to mention the complexity of the contexts in which practical mathematics developed. These difficulties are especially pronounced in the Orthodox world, whereas interest for practical mathematics began earlier and developed more strongly in Western Europe. But more in-depth studies concerning Greece, the Balkans and Rus' would be needed in order to draw an accurate picture (cf. Karp and Schubring 2014: 143–9).

Elementary learning began with Reading, followed by wRiting and finally Reckoning (the "three Rs"). As an example, a manual for English schoolteachers, dated 1587, centered the curriculum on reading in English, along with writing, numbering "by letters and figures" and casting accounts. Yet, there was no uniformity, even in comparable social environments, as scattered evidence shows us. In 1460, a merchant from Lyons—a great financial and commercial center at that time—offered advice to his son through a poem, known as *La complainte de François Garin* (The Lament of François Garin):

> when you finish your education at school, [. . .], if you want to join the merchants, you must not lose time reading nice stories and books, but learn to count because this is the way to understanding gold and currency accounts. (*La complainte de François Garin*, Lyon 1978: 17)

At the other extreme, a little later in the sixteenth century, we learn of a pastel maker from Lauragais near Toulouse, a successful businessman who could neither read nor write. Recorde regrets the "contempt or mysregarde" of the English for knowledge. Moreover, Roger Ascham (1515–68), tutor to

Princess Elizabeth, warned that unless it was taught in moderation, mathematics overcharged the memory.

Private tuition remained the more common way of educating nobles and the children of the rich bourgeoisie, whereas less wealthy parents were reluctant to school their children since they needed the income from their labor. Children who received vocational training often gained it on-the-job, whether working in a shop, workshop or office. This applied as well to the youngster who became the chief bookkeeper in the very powerful Fuggers' firm of Augsburg, which engaged in mining, international finance, and trade. He started his training while working at his father's business in Augsburg, then spent some years studying under a master in Venice. After returning to Augsburg, he improved his knowledge before writing a manuscript on bookkeeping.

An academic study on apprenticeship in Antwerp from the fifteenth to the end of the eighteenth century shows that before the advent of a formal schooling system, training and education "under the roof of a master" constituted a very flexible and dynamic system, perfectly in tune with the demands of an early modern economy. Many of those who wanted to learn mathematics studied them on their own. John Wallis (1616–1703) himself, who later became Savilian Professor of Geometry in Oxford, learned practical arithmetic from books during holidays, with the help from a young brother (Thomas 1987: 110; Denniss 2011: 455). Many authors of arithmetic textbooks claimed that they enabled a person to learn the subject "without a master." Some opted for a presentation based on a dialogue between Master and Scholar, a popular style from this period.

Public schooling depended on the local or national context. By the end of the Middle Ages, a number of elementary schools taught the "three Rs" (along with Latin fundamentals in the so-called Latin schools). These schools flourished in major European cities, though the level of education was entirely dependent on the master's knowledge. Provided he was of good moral character and could read, write, and count, nothing more was required to head an elementary school. Children often left just after learning to read. Moreover, their teachers were not always able to provide mathematics lessons beyond addition. Schools in rural areas often opened and closed in quick succession, as many teaching masters led an itinerant life, whereas in cities they were usually well established. In Genoa in Italy, elementary education had been organized as early as the thirteenth century; the city's businessmen typically sent their children to private or communal schools. In Pamiers, a small but active town located in the South of France, home schooling remained the exception. Up until 1540, merchants, lawyers, and notaries entrusted their children to the local municipal school. In Germany, up to Luther's time, schools in which children were taught to read and write the vernacular were few in number and poor in quality.

Religious and moral training was one of the main engines behind the interest in the development of schools. Education valued not only as a way of shaping minds, it was seen as a remedy for idleness and vagrancy, a factor in strengthening faith, morals and social order. Religious reformers played a major role in popularizing education, regardless of gender and social status. Philipp Melanchthon (1497–1560)—who studied the mathematical sciences—was the principal constructor of the Protestant educational system, whose responsibility was the affair of princes and magistrates. The Catholic Church actively promoted this movement for education by founding parish or urban schools under the supervision of priests and the elites who financed the recruitment of teachers. By the end of the seventeenth century, the majority of boys' schools were run by vicars or priests.

Schooling carried costs that poorer families could not afford, though several countries created charity schools. Some hospitals, such as the Grey Friars in London, provided education for poor or orphaned children who learned to read, write, and count, though some found this a senseless undertaking. In an essay from 1681, the writer asked why "must a poor Boy that is destined for a Mason, Bricklayer, Shoemaker [. . .] be taught [. . .] so far in Arithmetic, as if he were destined for a Merchant?" (Thomas 1987: 111). By the second half of the sixteenth century, urban charitable schools had become fairly widespread in France. However, schooling for the masses was not yet on the public agenda, and the financing for education remained dependent on the generosity of private individuals and local authorities. Thus, in 1666, admonitions were addressed to various authorities in the city of Lyons, noting that essential education of poor children was being neglected. Toward the end of our period, the French Catholic priest Jean-Baptiste de la Salle (1651–1719) dedicated much of his life to the Christian education of the "children of artisans and the poor." He founded the *Institute of the Brothers of the Christian Schools*, which spread to other countries, and promoted many enduring educational practices. What we know about the programs of these institutes certainly indicated that they must have been successful if they were actually implemented. After reaching the level of reading, pupils spent two half hours per week learning arithmetic: adding for two months, subtracting the next two months, then multiplying (three months) and dividing with proof (four months). Then came the rule of three, the divisors of a number and computations with fractions. This set of knowledge was more than enough for ordinary needs, either private or professional.

Most girls did not receive a high level of education, although there were notable exceptions among the social elites. Despite the emancipation movements which emerged in humanist Europe and the active participation of some women in running successful salons, the prevailing idea throughout this period was that a woman who took more than a passing interest in science looked ridiculous. The education of girls thus aimed to prepare them for their future educational and household duties. According to the abbot and lawyer Claude Fleury (1640–1723), who tutored several sons from families of high nobility, the girls would learn, *inter alia*, reading and writing, practical arithmetic and especially "economic" topics, in order to fulfil their tasks as well as possible (*Traité du choix et de la méthode des études* (Treatise on the Choice and Method of Studies, 1686)). The gap between boys' and girls' numerical education was not homogeneous in Europe but decreased everywhere over the years—the pioneering statistician William Petty (1623–87) declared in 1685 that "one day Arithmetick and Accountantship will adorn a young woman better than a suit of ribbands" (Wardhaugh 2012a: 75).

The Reformation's educational policies promoted girls' instruction as well as boys', whenever possible within the family. In 1530, a girls' school opened in Wittenberg, whereas in Geneva public schools offered free elementary education to boys and girls. In seventeenth-century England, some grammar schools accepted the presence of girls among the boys, though these cases were infrequent, as the gender mix was thought to be dangerous for girls' chastity. As early as the sixteenth century, schools for poor orphan girls were set up, providing Christian education and elementary knowledge. One of the best known in Europe was the Ursuline school, first founded in 1535 in Brescia, Italy. Beginning in the 1620s, girls from wealthier classes could pay to be educated inside the Convent, while poorer ones received free elementary education outside it. This model was adopted by other congregations and became popular and successful. In England, after the suppression of convents by King Henry VIII, secular boarding schools multiplied, each with its own clientele (noblemen, merchants' daughters, etc.). The teaching program, taught in the vulgar language, accorded with the families' expectations: reading, writing, sewing, music, arts, etc. In the seventeenth-century Ursuline schools were included reading in Latin and French, writing, as well as counting and reckoning with tokens and pen. After leaving the Ursuline convent, 12-year-old Marguerite de Bramereau wrote the *Rudiment d'arithmétique* (1655), from which we learn that she had been taught the four operations and the rule of three (Muffato 2018). In reality, girls outside the social elites often left school before being able to count.

After learning the basics of calculation, young people who planned to practice a job requiring some mathematical skills had to improve their knowledge. Were there any institutional structures available to meet those needs? A partial answer can be given in the case of commercial and financial training, as a corpus of textbooks has survived that bears witness to how this demand was met.

By the end of the Middle Ages, a new class of masters arose in answer to the increasing mathematical needs in commercial and financial trade. A new genre of mathematical work gradually emerged, whose success is attested by the fact that the first mathematical text to be printed, *Arte dell' Abbaco* (Treviso, 1478) belonged to this genre. Similar works soon spread throughout the great commercial and financial cities of Italy. Since church-sponsored schools were not well adapted to the growing needs of the business professions, private teachers, guilds or municipalities opened schools for this purpose, which expanded during the fourteenth and fifteenth centuries. Children between about ten and twelve years of age attended these schools, once they had learned to read and write. After about two years, most of them were ready for on-the-job training in a merchant's or craftsman's shop or in a banking office, whereas a few extended their knowledge by going abroad so as to enter international banking and commerce. Many masters began as commercial traders and then opened a school in their shop, hence the name *Botteghe d'ab(b)aco* (abacus shops). To support this teaching, hundreds of *Ab(b)aco* treatises were composed between 1400 and 1600, which, despite their name—here *abacus* is synonymous with calculation—were *algorisms*. Most of them are written in vernacular languages for a wider readership. Actually, the Italian "abacus movement" was part of a much broader intellectual movement that floated on the humanist wave. It attracted not only merchants but also architects and artists: for instance, the famous painter Piero della Francesca (*c.* 1415–92) wrote an *Abaco* treatise.

Elsewhere in Europe, where expansion of trade and finance was not so early nor so strong, the same phenomenon arose later. The would-be merchants from southern Germany, for instance, first had to train in northern Italy (Venice was particularly famous for training accountants) or learn firsthand in the accounting house of a German commercial establishment. By the end of the fifteenth century, we have the first evidence of a *Rechenmeister* (reckonmaster), Ulrich Wagner (?–*c.* 1490), in Nuremberg. During the sixteenth century, such masters established themselves in many larger cities, where they opened writing and reckoning schools (*Schreib-und-Rechenschule*). In 1613, the competition between the forty-eight teachers of Nuremberg had become so strong that they received an order not to establish schools too close to one another. The reckonmasters were valued and honored for their specific abilities. Adam Riese is still called on in Germany to testify to the correctness of a computation: "$3 \times 8 = 24$ according to Adam Riese." He had two chief professional functions: he ran a famous school in Annaberg and, in addition, he kept the records of the receipts and expenditures in the mining office of that town in Saxony.

The Netherlands followed the German model by founding schools that met the direct and indirect requirements of commerce and related needs in the growing cities, as Antwerp became the greatest commercial and financial center during the sixteenth century. These new institutions taught practical mathematics, together with such other topics as cartography, astronomy, or surveying. Private schools run by a *Rekenmeester* offered instruction in computation, accounting, principles of nautics, etc.

As for southern France and the Iberian Peninsula, these regions benefited from their proximity to Italy, as testified by manuscripts and books on commercial practices mostly from the second part of the fifteenth century. The influence of Italian *Abaco* treatises is evident, though not unique. What remains less clear is whether there were also schools on the model of the *Abaco* ones. We get a few indications from the authors themselves. Thus, Francesc Santcliment (15th century), who in 1482 published the *Summa de l'art d'Aritmètica* (Compendium of the Arithmetic Art), composed this work in the "remarkable city of Barcelona where he was teaching," though without giving any more detail. Nicolas Chuquet, the best known French mathematician of the late fifteenth century, wrote and taught commercial arithmetic. He was registered in Lyons as a writer (escripvain), where he taught writing and most probably gave lessons of *Algorism*.

Practical Arithmetics, originally emphasizing commercial applications, shared a common central core. They first cover a range of knowledge aimed at a broad readership: description of the Hindu-Arabic system of numeration, and/or (especially in northern Europe) "reckoning on the lines" with counters, exercises involving the four arithmetical operations, first with whole numbers, then with fractions. A second part deals with commercial rules, starting with the fundamental rule of three and others derived from it, in order to solve specific problems. All this is applied to concrete examples, involving units of currency, measurement, or weight. Other common topics are the extraction of square and cubic roots, as well as arithmetic and geometric progressions. A chapter on practical geometry (how to calculate areas and volumes, to measure lengths with instruments, with applications to surveying) is often added. Treatises exclusively devoted to commercial matters would skip topics, like the extraction of roots, which are of no use to merchants, but include, on the other hand, non-mathematical material, like commercial advice, lists of currencies in different areas or metallurgical techniques for the production of coins. Those textbooks that were pared down to the most essential practical needs were often among the most popular. They explained the rules by a set of instructions applied to a large range of examples, without any proof, and learned by rote. The training thus relied on repetition: readers and pupils were expected to recognize the type of problem they encountered, so as to apply the proper rule. However, these practices evolved over time and some authors warned against over-reliance on rules.

In the sixteenth century, disdain for practical arithmetic was still deeply entrenched in some minds. Its close ties with business—an activity long considered incompatible with nobility and the social elite—reinforced this attitude. The French sixteenth-century diplomat Blaise de Vigenère (1523–96) once wrote: "one must not study the divinely beautiful proportions of numbers in order to make them serve the computations of a bank" (Zemon Davis 1958: 3). Yet, as noted above, these cultural and educational changes reflected the growing economic demands and the new scientific requirements of astronomy, cartography, seafaring, etc. Throughout the century, the target audience of practical Arithmetics became much broader, while the views of the discipline evolved and the authors' social origins became more diverse.

A number of well-known scholars contributed to making practical mathematics a part of academic education, including Frisius in the Netherlands. He sought to bring together theory and practical or technical applications, constantly appealing in his *Arithmetica* to the "common uses," while openly relying on the Greek theory of numbers. Frisius's work is an illustration of the profound changes that were to come. In France, around the same time, debates and struggles took place between advocates of a practical training and the supporters of speculative mathematics. Among others, the great humanist Pierre Ramus (1515–72), one of the *Professeurs* at the *Collège royal* in Paris created in 1532, became a defender of practical mathematics as a result of his desire to reform the *quadrivium*. Ramus claimed that pupils must begin with counting and measuring and not with "belles abstractions." He stressed that practical rules, especially those of commerce, do not pervert pure arithmetic; these works benefit everyone, for the theorist would learn from practice and the practitioner from theory. Ramus highlighted the importance of arithmetic in daily life, in commerce and craft.

Practical Arithmetics were written in Latin or in vernacular languages, depending on their audience. However, Latin was losing its hegemony, even if slowly, whereas books written in a mother tongue were accessible to a larger audience. Jacques Peletier du Mans (1517–82), a French humanist, poet and mathematician, chose to write an *Aritmetique* on a high mathematical level in French. His argument was that writing mathematics—a discipline where the truth is infallible and constant—in French would enable the language to conquer the world of scholarly nobility. But Peletier progressively returned to Latin to please scholars, while others wrote reluctantly in French to meet the expectations of practitioners.

NUMERACY IN DAILY LIFE

It is difficult to evaluate how widely numeracy was spread in the population at large, a systematic study having yet to be undertaken in this field. The interest in numbers depended very much on a person's station in life, whereas concern for precise measurement was restricted to a limited sphere. Thus, for many, it was unnecessary to know one's exact age or to measure time accurately. Information related to domestic life is scattered through various types of archives and often caught only by chance. For example, we learn from an English schoolmaster of the early seventeenth century that some people were so ignorant that they were unable to turn to the proper chapters called out in a church service (Thomas 1987: 118). With regard to professional use, information is not so elusive. Apart from individual testimonials, which are quite rare, specialized works are available, such as trade practice books. These provided valuable information on currencies or prices, and sometimes included arithmetical explanations; carpenter's pocketbooks might also contain calculations related to construction, etc. For the period up until the end of the sixteenth century, practical Arithmetics, as said before, offer a major source of knowledge about needs and practices, as they allow us to identify some pressures that led certain classes of people to acquire new numerical skills. Nonetheless, we must remain cautious: first, because focusing on the educational aim blurs the boundary between reality and fiction; and second, the intended audience differs from the actual one. However, we can find quite reliable evidence of textbooks written by field practitioners, who worked as surveyors, employees in a mint, accountants, or financial experts. The content of such works (or at least the bulk of it) reflects the author's experiences, which makes it reasonable to assume that they were written as manuals for training would-be practitioners. We will be careful to give priority to these works.

It has long been assumed that the main audience for arithmetical textbooks were merchants. As John Dee wrote in his *Mathematicall praeface* to Henry Billingsley's English translation of Euclid's *Elements* (1570): "Who, nearer at hand, can be a better witness of the fruit received by Arithmetic, than all kinds of Merchants? Though not all, alike, either need it, or use it" (Dee 1570: iij). The author of an introduction to a book on accounting from 1588, advises "every good and discreet merchant" to be "prompt and readie in his reckoning and account making," and to be able to calculate "with penne or compters" (Thomas 1987: 106). Indeed, the word "merchant" refers to a very large group of people, from shopkeepers to men involved in large-scale international trade, from retailers to wholesalers. All of them had to make calculations, but commercial arithmetic was chiefly aimed at those whose business required mathematical knowledge beyond the four basic arithmetical operations, for a lack of numerical knowledge could bring financial disaster. Numerous sources testify to the increasing need for mathematics in other professions. Thus, Trenchant's popular *Arithmetique* was re-edited about twenty times. In the 1602 edition, a great range of professionals appear: merchants, elected officials, receivers, treasurers, masters of the Mint, "great lords who are in charge of republics, captains and warlords" (Trenchant 1602: A_2v^o). An interesting illustration of the mathematical skills needed at the Mint comes from the story of John Reynolds, an Assay Master (Biggs 2019), who was fourteen years old in 1599 when he started his ten-year apprenticeship at the Mint in the Tower of London. Reynolds was gifted in arithmetic, a valuable asset for his work in assaying and refining metals, an important part of which involved making silver and gold coins. The value of a coin depended on its weight and purity (fineness), which fluctuated and differed across countries. Every task required extreme precision and involved complex systems of measurement, along with complicated calculations. To streamline this work, Reynolds compiled several tables, which required a high level of expertise. Other contemporary works written by practitioners (goldsmiths, gaugers, etc.) testify to an increasing reliance on numbers in a wide range of professions, as reflected in more and more arithmetic textbooks. For instance, the introduction of excise taxes on alcohol in England (1643) stimulated the demand for measuring the contents of irregularly shaped barrels. Reynolds was also involved in these gauging problems. John Darling, in the fourth edition of *The Carpenter's Rule Made Easy* (1658), or "the art of measuring Superficies and Solids, as Timber, Stone, etc." adds an appendix "for gauging and making of casks."

The Frisian master and treasurer Edouard Leon Mellema (1544–1622) dedicated the second edition of his *Arithmetique* to the Count of Nassau. According to Mellema, without the help of his book the Count would not have been able to mediate disputes, share inheritances, ensure justice and the rights of everyone, pay the salaries of people on horseback and infantry, etc. Others mentioned as potential readers craftsmen, seamen, lawmen or surveyors, with chapters dedicated to practical geometry and surveying, since arithmetic was a prerequisite for the measurement or the division of lands. Jean Abraham, who, apart from being a French teacher of mathematics, was also a surveyor, imparted a good deal of practical advice in his *Arithmetique et arpentage universel* (*Universal Arithmetic and Surveying*, 1607). He emphasized that a good surveyor had to juggle the different regional units of measurement, to manage with direct or remote measuring instruments and to know how to calculate areas of land.

During the seventeenth century, a new form of rationality emerged, and numbers took on an increasingly important role in society and administration. Data collection relating to population census, tables for life insurance, etc., opened the way to demography and statistics. Some protagonists of these new ventures were originally craftsmen, or merchants, who were not necessarily highly educated in mathematical sciences (Cohen 2005: 45–58).

Most daily-life situations involving numbers, however, do not require advanced mathematics. The four basic operations and eventually the rule of three were usually sufficient, though even these skills were not widely shared. A popular children's rhyme from the late sixteenth century underscores that point: "Multiplication is mie vexation / Division is quite as bad / The golden Rule is mie stumbling stool / And Praktice makes mie bad" (Thomas 1987: 115). As noted earlier, adding up many numbers was often considered easier and safer when performed with tokens. Counting boards were frequently mentioned in household inventories, a testimony to their lasting use, notably by women. The celebrated French letter-writer Madame de Sévigné (1626–96) confirmed this practice when she wrote to her daughter on June 10th, 1671: "With these good counters, we have found that my property amounts to 530,000 livres, counting all my small inheritances." Even in the eighteenth century, the ability to "cast counters" was sometimes considered an important qualification for a marriageable girl.

Anyone who wanted to count one's money or to measure or weigh something was faced with the difficulties presented by the various metrological systems, due in particular to the irregularity of the ratios between the subunits. Their great diversity made the communication, trade and commerce between even nearby regions exceedingly difficult. The following example, concerning currencies in use in Paris during the sixteenth century, illustrates the types of problems that occurred. One *livre parisis* was worth 20 *sous parisis* which had the value of 25 *sous tournois*. On the other hand, 1 *sou parisis* was worth 12 *deniers parisis*, which were equivalent to 15 *deniers tournois*. In the Parliament of Paris, accounts were posted in *livres* and *sous parisis*, but *livres* and *sous tournois* were used in the *Chambre des comptes*. The numerical ratios between units and subunits of the same kind led to a number of specific reckoning methods to facilitate calculations, as, for example, in the currency system "*livre, sol, denier.*" Suppose 1 *aune* of linen costs 6d, how much will 18 *aunes* cost? Instead of multiplying 6 with 18 to get the value in deniers and then dividing by 12 to obtain the cost in *sols* and *deniers* (1 *sol* is worth 12 *deniers*), it was customary to first divide 18 by 2—since 6 is the half of 12—to get the result directly in *sols*. If the cost had been 5d instead of 6d, which is not so convenient, one would decompose 5 into 4 (the third of 12) and 1 (the twelfth of 12). A lot of these calculating tricks involving money or weights, presented as "brief and special" rules, were certainly very popular and spread in professional and perhaps also in daily-life calculations. They were especially useful to financial officers who constantly handled money and can be found in books used by moneychangers, who knew these rules by heart. In his routine work of minting coins, John Reynolds also had to make extremely complex calculations in such unwieldy metrological systems (a pound of weight was divided in seven subunits—ounce, pennyweight, grain, mite, droits, periot, blank—with the alternative ratios of 20 or 24).

At the turn of the seventeenth century, the eminent Flemish scientist Simon Stevin (1548–1620) attempted to introduce a simplified system using base ten arithmetic. Stevin, who made noteworthy contributions to science, technology, and engineering, "reinvented" decimal fractions with a proper notation in *De thiende* (The Art of Tenths, 1585). This work advocated a decimalization of all systems of coinage, measure, and weight for purposes of simplifying human accounts and measurements, taking examples from surveying, cask gauging, trade, etc. Another text from 1619, *The Art of Tens or Decimall Arithmeticke*, also advocated decimals in order to avoid the "intricacie of fractions" (Thomas 1987: 116). Nevertheless, though this numerical innovation was adopted here and there, Stevin's plea was not widely heard before the end of the eighteenth century ("a kind of tenaciousness of old customs," as underlined by Wallis).

Lacking decimal numbers, the result of any division was given as an integer plus a fraction, which is why calculations with fractions—perceived as difficult—

occupied such an important place in arithmetic. Fractions appeared as well in official calculations, like the *quarto y vintena* rule in Portugal. In the sixteenth century, when ships docked at the port of Lisbon, every cargo was checked in the *Casa da India* and submitted to taxation, to be paid to the king. The *Casa da India* was founded in 1503 under the reign of Dom Manuel I to manage international trade with the East (mainly spices); it asserted monopoly control on the transit of goods in favor of the king. *Quarto y vintena* means that a quarter was first taken off the price, the remaining money was then divided by 20, and the result was added to the first quarter. In our notations, the tax calculated by the employees of the *Casa da India* was $1/4 + 1/20 \times 3/4 = 23/80$ of the initial amount. Why such a complicated calculation, which appears like two successive tax collections? To determine the tax, the process was always the following. Starting from 80 is convenient, because one quarter is 20, three quarters are 60 and 3 is one twentieth of 60. 80 therefore leads to a tax of 23. Now, for any other amount, the rule of three must be applied. For instance, what will be the tax on 146 quintals of pepper arriving in Lisbon? If 80 quintals yields a tax of 23, 146 quintals will be taxed at "41 *quintaes*, 3 *arrobas*, 28 *arrateis* and 11 *onças*" (Ruy Mendes, *Pratica d'arismetica*, Lisboa, 1540).

Subunits avoid fractional results and mean more to users. There is also evidence that people did not use fractions fluently in everyday calculations. For example, some merchants' ledgers, at least up to the 1550s, commonly used only basic fractions, such as halves, thirds, quarters or tenths. The division of units into many subunits together with special reckoning rules, such as the one described above, made it possible to overcome these sorts of difficulty. Thus, a tailor from Paris published a booklet entitled *Le tailleur sincère* (The Honest Tailor) in 1671 to teach apprentices in this trade how to calculate the quantity of fabric required to make a garment with as little waste as possible. In his description of fractional lengths, the expression "half of a quarter"—a quarter being more like a subunit than a fraction—comes up frequently, along with only a few other fractions like the "twenty-fourth" or "thirty-second part," always written in letters, while Arabic figures are used in the clothing patterns.

The rule of three, introduced above, is also called the golden or merchants' rule, or sometimes the rule of proportion. "The golden rule" highlights its excellence, "the merchants' rule" emphasized its chief role in all commercial transactions, and "the rule of proportion" is the scientific name: given three numbers, it shows how to calculate a fourth one so that the four quantities form a proportion. This is the proper rule Clown needed to apply to answer the question he asked himself, as described at the beginning of this chapter. In his *Arithmetique*, Jean Abraham asks: if 40 jars of wine are worth 28 pounds, how much would 58 jars be sold? This was not considered an easy problem, as most of the authors explained how to apply the rule in considerable detail by first identifying quantities of the same kind, then the spatial organization of the three numbers and finally making the calculations. The golden rule—and its variations—allowed people to solve all the most common problems, including conversions from one currency to another, or money fluctuations. It gave birth to many other useful rules that remained in use for centuries. One of the most famous and common is the "rule of fellowship." Companies or corporations were born in medieval Italy when merchants felt the need to create associations in order to increase their profits and reduce risks. Sharing money between partners necessitates exact calculations! Introducing the "rule of fellowship" in his 1648 pocketbook of arithmetic, written for those with no interest in theoretical knowledge, Jacques Savary (1622–90) notes that this rule is universally useful to all people in charge of finance in any administration (like tax collectors or treasurers), as well as to accountants and merchants. The simplest form of a corporation was an association for a given period of time, each member providing some start-up capital. When the partnership was dissolved, the profits (or losses) were shared in proportion to the contributions. The solution was obtained by applying the "rule of fellowship," based on the rule of three and learned by heart: dividing each contribution by the total capital and multiplying by the gain (or loss) gives the amount allocated to (or taken from) each partner. Associations were often more complicated, as partners did not always remain together during the same time period. Or sometimes a fellow might invest nothing but his labor, which had to be numerically assessed. Finally, there were cases when contracts could not be carried out as initially planned and contract breaches that had to be solved as fairly as possible. All these issues concerned rather important merchants.

Bartering was used by more people, especially when coins were scarce (the practice became less common in the seventeenth century). In barter exchanges, prices were always increased, as illustrated by the following example of the baseline situation. Two merchants enter into barter exchange, one for pepper and the other for fabric. The merchant with pepper asks for 25 *llivres* a quintal in the barter instead of 20 *llivres* in cash. A *cana* of cloth is worth 19 *sous* in cash. By how much should each *cana* be overpriced so that neither party is cheated? Francesc Santcliment (*Summa de l'art d'aritmètica*, Barcelona 1482) gives the answer: first find out how much pepper is oversold, then use the rule of three to calculate how much fabric must be oversold as well. Bartering becomes a real headache when part of the transaction is in ready money and part in barter. Deception was a frequently alleged practice in business life and barter was particularly famous for it. Mathematics was thus seen as a useful tool to prevent one party from deceiving the other, but chiefly to avoid being deceived.

Many other commercial and financial applications could be addressed—calculations of simple and compound interest, currency exchange rates, or proportions in metal alloys—all matters that are historically established as practices for which fixed rules were transmitted and automatically applied. Among the topics unrelated to money, we will focus on the military field. During the sixteenth century, the art of warfare moved into a more scientific phase, integrating natural philosophy, geometry, and arithmetic. The size of armies grew massively within half a century, and this growth contributed to tactical thinking based on calculation (Brioist 2015: 112). Many an author advocated the study of practical arithmetic and geometry for military purposes. Dee reported that the Earl of Warwick had a set of rules for organizing his armies (kinds of weapons, artillery, footmen, or horsemen, etc.), which he kept in a golden case he wore around his neck. This case was "his most precious jewel, and most trusty counsellor" (Brioist 2015: 115). Some textbooks discussed sophisticated methods of calculation, others produced tables to be memorized, mainly directed at people ignorant of the art of reckoning. The *Brèves tables* (Brief Tables) published by Girolamo Cattaneo (1540–84), re-edited many times after 1563, were translated into various languages. The tables give the solution to problems such as forming squares of pikemen with certain constraints, such as conditions on the proportion of men bearing specific kinds of weapons. In the mid-sixteenth century, the famous mathematician and engineer Niccolò Tartaglia (*c.* 1499–1557) was the principal mathematical reference in matters relating to military science. His book entitled *Quesiti et inventioni diverse* (Various Questions and Inventions, 1546) deals with varied topics such as the firing of artillery, cannonballs, the disposition of infantry, topographical surveying, and statics. One problem concerns how to arrange foot soldiers in a square when their number is not a perfect square. Tartaglia chooses the example of 35,000 men and begins by finding the square root, which turns out to be 187 with a remainder of 31 soldiers. Thus, if 187 men are placed in each line and row, the army will be arranged in a square shape. Regarding the remainder, Tartaglia adds: "the above-mentioned 31, the sergeant will accommodate them wherever he wants, but I say that this residue should rather be put at the back of the formation" (Brioist in van Damme 2015: 116). Apart from (and certainly before) learning the extraction of roots, which was not an easy task, officers were advised to master operations with fractions and to understand the rule of three, both useful in food management for regiments.

Mathematical knowledge was otherwise highly disparate. Diverse types of tables were published in the sixteenth and especially in the seventeenth century. At the back end of Estienne de la Roche's (1470–1530) *Arismetique* (Lyons, 1538) one finds ready-made accounting tables. These enabled easy tabulation for "purchases and sales of all goods, either wholesale or retail, by measure, or by weight, by load, or by number," including, among others, tables used by bookmakers, related to the price of paper. In 1533, Riese had published the

"Annaberg Bread tables," a "pre-computed" booklet with numerical tables for calculating everyday prices. According to the foreword, it was to help "the poor common man not be cheated when buying bread." In 1582, Stevin published in Flemish the first printed tables for calculations of interest (manuscript tables had been in use long before, but they were considered secret tools by their owners who refused to publish them).

The development of printing favored the creation and dissemination of books or booklets entirely devoted to tables, which poured out in profusion during the seventeenth century. François Barrême (1638–1703?), who was an expert at the *Chambre des comptes* (audit chamber) in Paris and had previously given many lessons on bookkeeping, wrote several *Ready reckoners* in the late seventeenth century. He entitled one of them "The book necessary for accountants, lawyers, notaries, prosecutors, traders and generally for all kinds of professions." It encloses all sorts of tables (change, discount, interests . . .). It is a precious tool to save time ("everything is done at a single glance") or help poorly numerate people. It ends with a tariff to divide amounts of money up to 30,000 pounds, "without having learned division."

During the seventeenth century, a growing divergence appeared between elementary textbooks and specialized literature, due to the increasing complexity of social, economic, and administrative practices and the parallel evolution of mathematics: compound interest, double-entry bookkeeping, risk calculation (can Arithmetic help when experience loses its rights?). To address issues such as these, tables were of great importance.

Traditionally, many Practical Arithmetics have also included recreational mathematics, which aimed to "sharpen the minds" with surprising and appealing problems. "Think of a number"—a long-standing problem—is an emblematic example of these riddles: ask someone to think of a number, then give him or her a succession of instructions: triple the number (add 1 if the result is odd), take the half, triple the result, tell how many times the factor 9 is included in the number, and for each 9, take 2. Add the two (if the first result was odd, add 1 to the number of two); this is the assumed number. What looks like a magic trick could be easily rationalized with elementary arithmetic. The above problem appears in Claude Gaspard Bachet de Méziriac's (1581–1638) *Problèmes plaisants & délectables qui se font par les nombres* (Pleasant and Delectable Problems that are Solved by Numbers, 1624). This work stands at the origin of a new genre devoted to mathematical recreations, which would soon spread throughout Europe. The first book actually bearing this name, openly "serviceable to all both for pleasure and recreation" was published in 1624 and translated into English in 1633 (H. van Etten, *alias* Jean Leurechon, *Mathematicall Recreations*). However, works like Bachet's were chiefly aimed at "ingenious spirits" rather than seeking to popularize mathematics for a broader population.

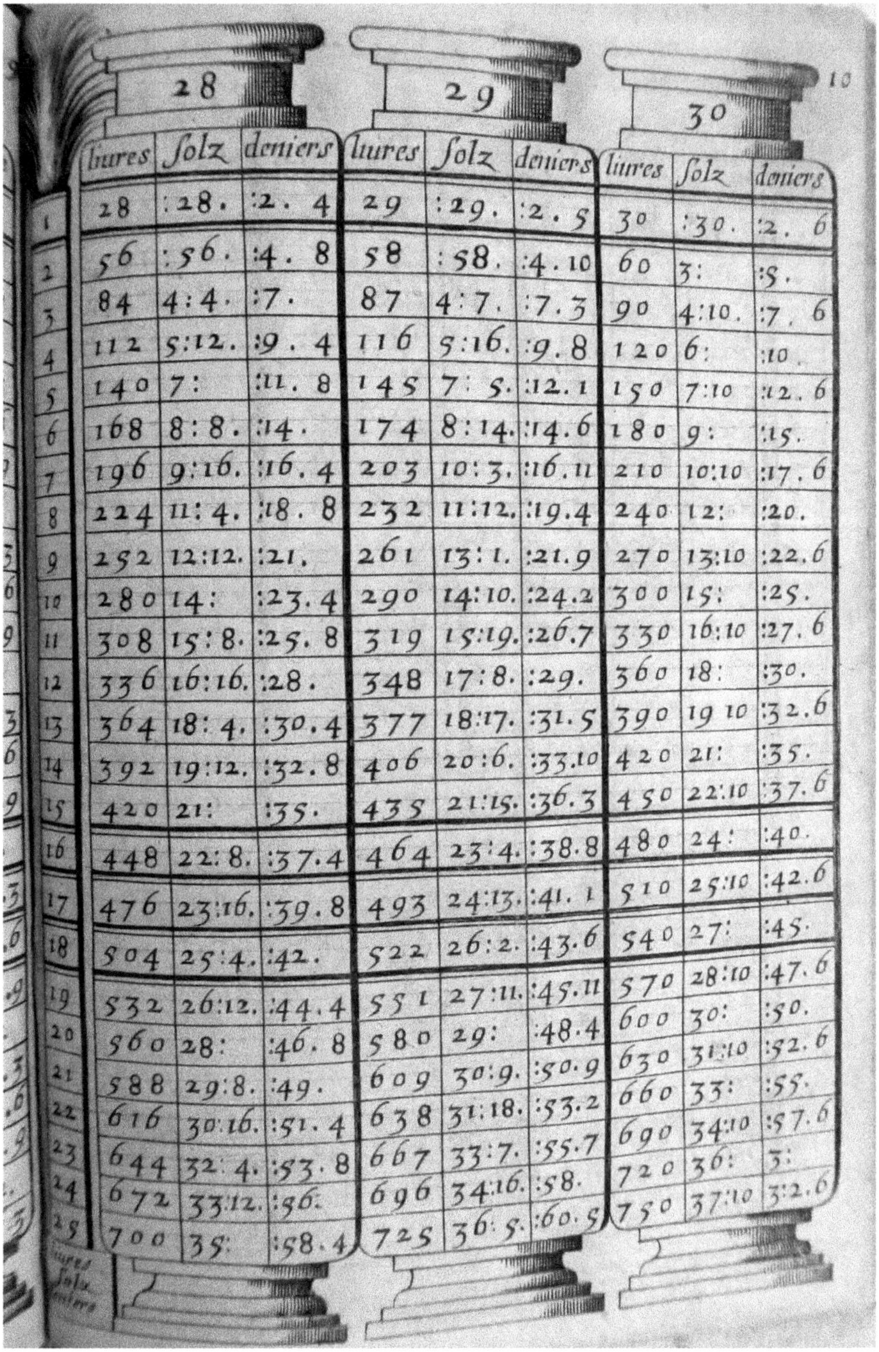

	28			29			30		
	livres	solz	deniers	livres	solz	deniers	livres	solz	deniers
1	28	:28.	:2. 4	29	:29.	:2. 5	30	:30.	:2. 6
2	56	:56.	:4. 8	58	:58.	:4. 10	60	3:	:5.
3	84	4:4.	:7.	87	4:7.	:7. 3	90	4:10.	:7. 6
4	112	5:12.	:9. 4	116	5:16.	:9. 8	120	6:	:10.
5	140	7:	:11. 8	145	7: 5.	:12. 1	150	7:10	:12. 6
6	168	8:8.	:14.	174	8:14.	:14. 6	180	9:	:15.
7	196	9:16.	:16. 4	203	10:3.	:16. 11	210	10:10	:17. 6
8	224	11:4.	:18. 8	232	11:12.	:19. 4	240	12:	:20.
9	252	12:12.	:21.	261	13:1.	:21. 9	270	13:10	:22. 6
10	280	14:	:23. 4	290	14:10.	:24. 2	300	15:	:25.
11	308	15:8.	:25. 8	319	15:19.	:26. 7	330	16:10	:27. 6
12	336	16:16.	:28.	348	17:8.	:29.	360	18:	:30.
13	364	18:4.	:30. 4	377	18:17.	:31. 5	390	19 10	:32. 6
14	392	19:12.	:32. 8	406	20:6.	:33. 10	420	21:	:35.
15	420	21:	:35.	435	21:15.	:36. 3	450	22:10	:37. 6
16	448	22:8.	:37. 4	464	23:4.	:38. 8	480	24:	:40.
17	476	23:16.	:39. 8	493	24:13.	:41. 1	510	25:10	:42. 6
18	504	25:4.	:42.	522	26:2.	:43. 6	540	27:	:45.
19	532	26:12.	:44. 4	551	27:11.	:45. 11	570	28:10	:47. 6
20	560	28:	:46. 8	580	29:	:48. 4	600	30:	:50.
21	588	29:8.	:49.	609	30:9.	:50. 9	630	31:10	:52. 6
22	616	30:16.	:51. 4	638	31:18.	:53. 2	660	33:	:55.
23	644	32:4.	:53. 8	667	33:7.	:55. 7	690	34:10	:57. 6
24	672	33:12.	:56.	696	34:16.	:58.	720	36:	3:
25	700	35:	:58. 4	725	36: 5.	:60. 5	750	37:10	3:2. 6

FIGURE 1.6. Alexandre Jean 1649. *L'arithmétique au miroir . . .*: 10, kept in Toulouse, Bibliothèque d'étude et du patrimoine. Photo by Maryvonne Spiesser.

Not devoted to mathematics per se, almanacs contained numerical information (see, for instance, Figure 1.7) that was easily accessible for the reader (like reading the calendar). They began to appear in Europe at the end of the fifteenth century and soon attracted a wide readership, since they were cheap, small, and handy annual publications, also useful because they included necessarily the calendar for the coming year. Almanacs were often dedicated to specific places and served many crucial functions in people's lives. They gave information on weather forecasts, farmers' planting dates, tide tables, and provided practical, medical, and household advice. They also included a section on prognostication and revealed the supposed influences of the stars on human affairs—the authors often claimed to be professors of astrology and mathematics or physics. Many an almanac also gave some mathematical information in the form of ready reckoners—like tables of interest—and eventually provided monetary calculations. The development of almanacs was fueled by the advances in print technology. They became tremendously successful, with a large audience

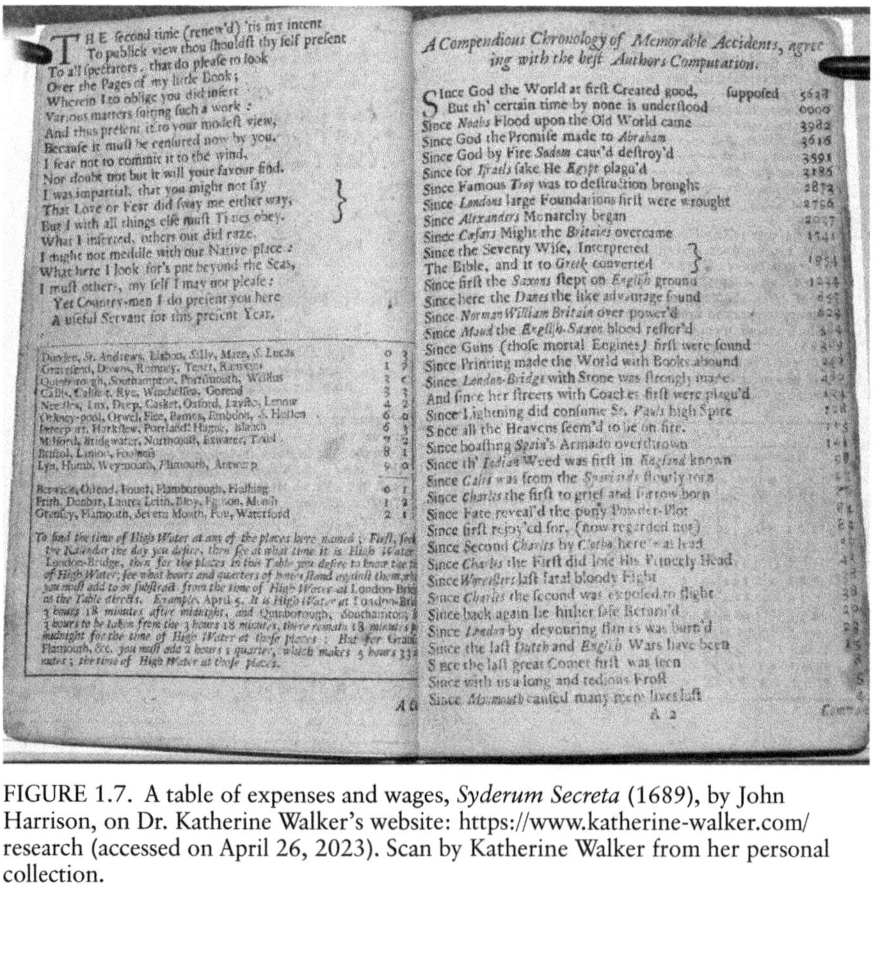
THE ſecond time (renew'd) 'tis my intent
To publick view thou ſhouldſt thy ſelf preſent
To all ſpectators, that do pleaſe to look
Over the Pages of my little Book;
Wherein I to oblige you did inſert
Various matters ſuiting ſuch a work:
And thus preſent it to your modeſt view,
Becauſe it muſt be cenſured now by you.
I fear not to commit it to the wind,
Nor doubt not but it will your favour find.
I was impartial, that you might not ſay
That Love or Fear did ſway me either way,
But I with all things elſe muſt Times obey.
What I inſerted, others out did raze,
I might not meddle with our Native place:
What here I look for's put beyond the Seas,
I muſt others, my ſelf I may not pleaſe:
Yet Country-men I do preſent you here
A uſeful Servant for this preſent Year.

A Compendious Chronology of Memorable Accidents, agreeing with the beſt Authors Computation.

Since God the World at firſt Created good, ſuppoſed 5638
But th' certain time by none is underſtood 0000
Since *Noahs* Flood upon the Old World came 3982
Since God the Promiſe made to *Abraham* 3616
Since God by Fire *Sodom* cauſ'd deſtroy'd 3591
Since for *Iſraels* ſake He *Egypt* plagu'd 3186
Since Famous *Troy* was to deſtruction brought 2872
Since *Londons* large Foundations firſt were wrought 2756
Since *Alexanders* Monarchy began 2037
Since *Cæſars* Might the *Britains* overcame
Since the Seventy Wiſe, Interpreted
The Bible, and it to *Greek* converted 1954
Since firſt the *Saxons* ſtept on *Engliſh* ground
Since here the *Danes* the like advantage found
Since *Norman William Britain* over power'd
Since *Maud* the *Engliſh-Saxon* blood reſtor'd
Since Guns (thoſe mortal Engines) firſt were found
Since Printing made the World with Books abound
Since *London-Bridge* with Stone was ſtrongly made
And ſince her ſtreets with Coaches firſt were plagu'd
Since Lightning did conſume *St. Pauls* high Spire
Since all the Heavens ſeem'd to be on fire.
Since boaſting *Spain*'s Armado overthrown
Since th' *Indian* Weed was firſt in *England* known
Since *Calis* was from the *Spaniards* ſtoutly torn
Since *Charles* the firſt to grief and ſorrow born
Since Fate reveal'd the puny Powder-Plot
Since firſt rejoyc'd for, (now regarded not)
Since Second *Charles* by *Cloths* here was lead
Since *Charles* the Firſt did loſe his Princely Head
Since *Worceſters* laſt fatal bloody Fight
Since *Charles* the ſecond was expoſed to flight
Since back again he hither ſafe Return'd
Since *London* by devouring flames was burn'd
Since the laſt *Dutch* and *Engliſh* Wars have been
Since the laſt great Comet firſt was ſeen
Since with us a long and tedious Froſt
Since *Monmouth* cauſed many mens lives loſt
A 2

FIGURE 1.7. A table of expenses and wages, *Syderum Secreta* (1689), by John Harrison, on Dr. Katherine Walker's website: https://www.katherine-walker.com/research (accessed on April 26, 2023). Scan by Katherine Walker from her personal collection.

(they were "profitable to all sorts of men") ranging from the members of high society to "forgotten working women and men for whom the almanac [. . .] might also be their only contact with the world of numbers and calculation" (Wardhaugh 2012a: 2).

Poor Robin, first published in 1663, was one of the most popular and long-lived almanacs in England. It "stood, on the whole, [. . .] for the fact that mathematics could be fun and that numbers could be a tool in the hands of the powerless as well as the powerful" (Wardhaugh 2012a: 6). In the crowded English market, one could find an almanac in every third English household by the 1660s.

CONCLUSION

The spread of popular numeracy cannot but be elusive. While numbers were increasingly essential to many professional fields, only a smattering of arithmetical competence was generally needed for daily private life. For people who came to arithmetic late in life, it appeared a difficult matter, apt to be forgotten. It was assumed that a gentleman was equipped with rudimentary mathematical education, but more than one had to find help to do simple accounts! There are many ways to deal with numbers, as the enduring use of tally sticks and counting tokens testifies. Among professionals, practical field experience and the rule of thumb were still often prioritized over book-based learning. We must therefore be cautious when observing the insights offered by educational documents, although the works close to the ground of human experience do attest to contemporary practices and needs. The fact that arithmetic books became increasingly widespread in early modern Europe, some enjoying considerable success and several re-editions, and the fact that their readership grew accordingly, highlights how there was a strongly felt need for arithmetic—more widely, for rationality—in areas where experience, judgment, or the attachment to traditions had prevailed before.

CHAPTER TWO

Practice and Profession

Range and Change in Mathematical Lives

JIM BENNETT

The Latin title *Mathematicus*, or its linguistic equivalents, might derive from an official position, office, or responsibility, whether at court, in a university or in a civil administration. Alternatively an individual might come to be recognized as a mathematician through activates undertaken outside any office or employment. Almost every adult member of society uses mathematics, while only a few are acknowledged as mathematicians. By reviewing a wide range of examples, this chapter offers an account of how mathematicians achieved and sustained their status in early-modern Europe. The transcendent claims of mathematicians notwithstanding, mathematical practice is located in time and place, and to affirm this we will begin with a particular time (May 1547), followed by a more extended account of activity in a particular place (Nuremberg). By invoking the more abstract concept of *locus*, we then consider types of places, such as observatories, printing houses, and colleges, that are found in many locations. Finally, generalizing further, we look at principles, tools, and methodologies that are common to mathematical practice across time and location.

MATHEMATICAL ACTIVITY IN TIME AND PLACE

A Meeting in Louvain, 1547

When the English mathematician John Dee (1527–1609) visited Louvain in 1547, as he developed an ambition to remedy his country's backwardness in mathematical sciences, his plan was to confer with "some learned men and chiefly mathematicians," naming specifically Gemma Frisius (1508–55), Gerard Mercator (1512–94), Gaspar à Mirica (*c.* 1496–after 1549), and Antonio Gogava (1529–69) (Dee 1851). This is an instructive list. At its head is the instrument and globe designer and maker, Gemma Frisius, who has signed his only surviving globe, made in collaboration with Gaspar à Mirica and Gerard Mercator, as "medicus ac mathematicus." Gemma's formal qualification was in medicine, his university appointment was in the medical faculty and he was a practicing physician; his claim to *mathematicus* rested on his published works, his designs of instruments and his direction of a workshop producing instruments and globes. His portrait (see Figure 2.1) presents all of this with deliberation. In front of books and instruments, this is literally a very "hands-on" mathematician.

He grasps the celestial globe as a material creation—no symbolic allusion to the heavens here—while the table is littered with instruments for construction and drawing, and even scissors for cutting out the paper sections or "gores" for his globe. An instrument of his own design—the astronomer's rings—also rests on the table, where the urine flask speaks of Gemma's other profession as *medicus*.

What of the other mathematicians on Dee's list? Gemma's branch of mathematics would usually have been called "cosmography" in the sixteenth century (the study of the heavens and the earth together and their association through location, time, season and climate) and the same was true of Mercator. He had been a student at Louvain, undertaking serious study of mathematics only after graduation, under the direction of the cartographer Franciscus Monachus (*c.* 1490–1565) and particularly with Gemma himself. Associated with Gemma's workshop, Mercator made mathematical instruments, such as astrolabes, as well as globes. His engagement with mathematics was a serious one—he read the *De revolutionibus orbium coelestium* (On the Revolutions of the Heavenly Spheres) of Copernicus (1473–1543) and he is best remembered for his eponymous map projection tailored to the needs of navigators. He developed an independent business in globes and maps, and later became cosmographer to Duke Wilhelm of Cleves (1516–92).

Mercator had been trained to engrave by the third of Dee's mathematicians, the goldsmith and engraver Gaspar à Mirica or Gaspard van der Heyden. Like Mercator he collaborated with Monachus and with Gemma, and was a maker of instruments and globes. We have no record of a university education and an apprenticeship would have been the natural route for a master goldsmith, but he was treated as a mathematical practitioner and moved in humanist circles in Louvain.

Finally Antonio Gogava was yet another mathematical pupil of Gemma's, but his specialism was to be music alongside astrology. He translated Ptolemy's astrological work *Tetrabiblos*, to which he added a treatise on the parabolic

FIGURE 2.1. Printed portrait of Gemma Frisius, by Jan van Stalburch, 1557. Courtesy of Rijksmuseum.

conic section that he attributed (incorrectly) to Apollonius. It was published in 1548 with a preface by Gemma.

We must not forget that Dee himself was part of this mathematical encounter. In his famous "Mathematical Praeface" to the first English translation of Euclid's *Elements* of 1570 he set out and explained all the mathematical arts and sciences. In dealing with astrology, he recalled again his visits to Louvain and his "earnest disputations, of the Learned Gerardus Mercator, and Antonius Gogaua, (and other,)" (Dee [1570] 1975). Dee's mathematical training seems not to have stemmed from his education at Cambridge but owed much more to his time in Louvain, especially a second visit in 1549–50. Although he was offered, he said, a professorship of mathematics at the Universities of Paris, his mathematical career was not academic: he had income from clerical appointments, he provided mathematical services to noblemen, to trading companies and to the queen, including navigation and astrology, he taught mathematics and he published. His main base was his extensive private library.

It might seem that we have fallen in among an oddly miscellaneous and eccentric collection of mathematicians. None seems to have had a formal, institutional training in mathematics; none seems to have followed a recognizable career path provided by educational foundations or official bodies; each seems to have been living on his wits by taking opportunities as they came along and by making his own way. Much of their work may seem less mathematical than we might have expected, involving instrument production, entrepreneurial activity, and practical consultancy. Although there were contemporary instances of more settled and predictable employment, as we shall see, the variety in the career trajectories followed by these mathematicians before and after their encounter in Louvain, is not unusual.

Nuremberg c. 1530 as a Mathematical Intersection

Mercator was not the only mathematical cosmographer making globes and reading Copernicus. Johann Schöner (1477–1547) was a Catholic priest in the early years of the sixteenth century in Bamberg, Germany, where he had a printing press and produced maps and globes. He moved to the more active center for practical mathematics, Nuremberg, where he was appointed lecturer in mathematics in the school founded by the Lutheran scholar and educational reformer Philipp Melanchthon (1497–1560). Schöner became a protestant and taught there from 1526, while continuing to develop his globes. He produced the first printed globe and the first matched pair of celestial and terrestrial globes, reflecting the duality of the discipline of cosmography. A posthumous engraved portrait by Theodore de Bry (1528–98) places him in a monumental arch bearing only his name and the identity *Mathematicus*.

Schöner had a considerable reputation, and a pupil of Melanchthon and mathematics professor from the Lutheran stronghold of Wittenberg, Georg Joachim Rheticus (1514–74), spent time with him in Nuremberg. Schöner persuaded him to travel north to Frauenberg, where it was known that a canon of the cathedral, Nicolaus Copernicus, had a novel astronomical theory. Rheticus carried a letter of introduction from Schöner and among the books he took as presents was Schöner's edition of the systematic trigonometrical work, *De triangulis*, by the late Nuremberg mathematician and astronomer Johannes Regiomontanus (1436–76). The first published account of the Copernican heliocentric system was the *Narratio Prima* (First Account) written by Rheticus by agreement with Copernicus, published in 1540 and addressed to Schöner. When Rheticus returned with the precious manuscript of *De revolutionibus orbium coelestium* (On the Revolutions of the Heavenly Spheres), he went first to Wittenberg, but then brought it to Schöner, who helped arrange for its publication in Nuremberg.

Another Nuremberg mathematician who encouraged Rheticus to visit Copernicus was Georg Hartmann (1489–1564). He introduces a different emphasis to this network of mathematicians, for he was principally engaged in designing, making, and selling mathematical instruments, both in brass and printed on paper. The term "mathematical instrument" in this context needs explanation. These were not scientific instruments in the modern sense: they were not intended for investigation, experiment, or discovery. They helped with problems that could be treated with mathematical techniques in fields such as astronomy, time-telling, surveying, navigation, and certain arts of war, such as ballistics, range-finding, and designing and laying out geometrical fortifications. Hartmann was famous for his astrolabes, but also made globes, sundials, quadrants, and gunnery instruments. All of this made him a mathematician in the understanding of the sixteenth century but this commercial instrument-maker was also a clergyman and a Latinist, who published on optics and astrology and prepared a tract (unpublished in his lifetime) on sundials and astrolabes.

Hartmann was a friend of the artist Albrecht Dürer (1471–1528) of Nuremberg who, in spite of a limited education, became a serious and significant mathematician, publishing works on geometry, on fortification and on proportion. In a commercial sense, Hartman's printed paper instruments were the equivalent of Dürer's famous engravings, allowing multiples to be taken from a single plate. Much of the characteristic enterprise of Nuremberg instrument making would fall within the broad understanding of mathematics at the time. After a somewhat itinerant life, living from university teaching and from private patrons, Regiomontanus himself chose to settle there in 1471, saying that a principal reason was the availability of instruments. He too established a printing press in Nuremberg.

While we are considering mathematicians in southern Germany in the early sixteenth century, we can add two further instructive instances for building a picture of mathematical practice. Peter Apian (Petrus Apianus, 1495–1552) would be thought of principally as a cosmographer, while working also in astronomy and cartography. When he was appointed to the University of Ingolstadt in 1527, his responsibilities were as mathematician and printer. This is a combination we have already seen in professional practice; here it is given an institutional status. Apian's printing included paper instruments in his books—both his

Cosmographicus liber (Cosmography Book, carried on in revised editions by Gemma Frisius) and his magnificent *Astronomicum Caesareum* (Astronomy of the Emperors) of 1540, which was essentially a complete "equatorium," an instrument for finding the positions of the planets in Ptolemaic astronomy.

Our second German instance beyond Nuremberg is Johannes Stöffler (1452–1531), a Catholic priest and maker of instruments, clocks, and globes, who was appointed to a new Chair of Mathematics and Astronomy at the University of Tübingen in 1507. Stöffler's most influential book was an exhaustive treatment of the astrolabe, *Elucidatio fabricae ususque astrolabii* (Explanation of the Construction and Use of the Astrolabe, Oppenheim 1513), a detailed account of, as the title indicates, how to make the instrument and then how to use it. It is very firmly a practical work, using all the possibilities of book production, such as fold-out tags and paper instruments, to guide the reader/maker through the task, step by step. This intersection of instruments, printing, and the practical engagement of technical scholarship is typical of sixteenth-century mathematics. Stöffler, too, has a printed portrait in the series by de Bry, surmounted by the single descriptor beside his name, "Mathematicus."

PRACTICE AND LOCUS

Printing Houses; Observatories and Universities

Having examined two cases where place played a significant role in the mathematical narrative—one where a specific encounter occurred, another a point of productive intersection over a number of years—we may consider more generally how places with particular functions have created mathematical priorities or were themselves created in response to such priorities. We have seen already that mathematicians often also carried other professional identities—as teachers, instrument makers, cosmographers, astronomers, cartographers, or surveyors. Other parallel identities may be less obvious: for example, humanist scholarship in the editing and publishing of texts was a vocation that could be given a mathematical character, while commentaries issued with edited texts opened possibilities for original work. Regiomontanus and Schöner would qualify as mathematical editors, among their other activities, though not all of their publications were classical in the humanist tradition. Looking farther afield, a prominent instance would be Federico Commandino (1506–75) of Urbino, another mathematician visited by Dee (in 1563), who trained and to some extent practiced in medicine. His preoccupation with editing classical mathematical texts was not incompatible with work on instrumentation, in Commandino's case the proportional compasses. Another important scholarly mathematician would be Francesco Maurolico (1494–1575), a priest in Messina, Sicily. Though a Benedictine, he was appointed

Professor of Mathematics in the Jesuit College there in 1569. While he engaged in aspects of mathematical practice, such as astronomy and fortification, a major achievement was the recovery of mathematical works from ancient Greece, published alongside Maurolico's commentaries and his own original work.

The textual ambitions of the mathematical editors were served by printers who developed a specialist interest in mathematical publication, though this would never become an exclusive preoccupation for them. These publishing houses included Erhard Ratdolt (1447–*c.* 1528) in Venice, Johannes Petreius (1497–1550) in Nuremberg and perhaps especially the Officina Plantiniana in Antwerp, where Christophe Plantin (1520–89), succeeded by his son-in-law Jan Moretus (1543–1610), traded under the suitably mathematical sign of "The Golden Compasses." Another Antwerp publisher, Roeland Bollaert (fl. 1527–9), had become an agent for the sale of globes by Schöner but, finding that demand exceeded supply, he encouraged Gemma Frisius to enter the trade. Gemma adopted a number of features from Schöner's practice, not least the pairing of celestial and terrestrial globes, adopted in turn by Mercator and by globe makers generally for several centuries. Publishers had demonstrable influence and *loci* such as these printing houses can be thought of as forms of institution, where cultures of specialist learning and technical resource were established and developed, becoming settled parts of the mathematical infrastructure. The existence of outlets for specialist printing was an important element in the development of mathematics.

Astronomy was another aspect of mathematical practice that required particular appointed and equipped spaces: if not full observatories, at least settings where instruments could be used to advantage. These could be linked to practices we have already encountered. Regiomontanus and his friend Bernhard Walther (1430–1504) established an observatory in Nuremberg, together with Regiomontanus's printing press. Dürer purchased Walther's house after his death, while Schöner acquired his and Regiomontanus's papers and continued their publication ambitions. As a successful merchant, Walther's background introduces another example of the diverse careers of contemporary mathematicians. We shall return to the astronomer and mathematician Tycho Brahe (1546–1601), but for now he provides a further instance of the practical combination of an observatory and printing press, in this case set on the Danish island of Hven.

Michael Maestlin (1550–1631) at the University of Tübingen is an example of an influential mathematical astronomer who did not travel far from his birthplace but found in a nearby university a *locus* that set the course of his intellectual life as a mathematician. There, he was taught by Philipp Apian (1531–89), the mathematician son of Peter. Maestlin became Professor of Mathematics at the University of Heidelberg in 1580 and from 1583 taught mathematics at Tübingen for forty-seven years. An early Copernican, he has a strong influence on Johannes Kepler (1571–1630), though it is indicative of the contemporary understanding of the role of mathematical astronomy that he never accepted his pupil's radical notion that astronomical theories should be physical as well as mathematical. So far as universities were concerned and the academic distinctions they respected, there was a different set of scholars engaged in physical explanation, working in a separate discipline, that of natural philosophy. This was the distinction Kepler found irrational and unacceptable, and not least incompatible with what he regarded as his devotional vocation—finding God's design in the cosmos.

We see that universities constituted *loci* where mathematical practice could be shaped in the sixteenth century. However, their influence was not uniform and depended, not only on influential scholars, administrators, and patrons, but also on intellectual and confessional traditions. Some universities had special interests on account of influential principals or powerful professors, though often the teaching for the mathematical subjects in the standard curriculum was dull, repetitive and of low esteem.

Under the influence of Melanchthon at Wittenberg, a characteristic attitude to this question of the relationship between mathematical and physical explanation was developed in a form of particular relevance to Copernican astronomy but bearing also on other branches of mathematics, so we shall return to this when considering general principles that worked across different areas of practice. In the "Wittenberg interpretation," as it has become known to historians, Copernican mathematical theory can be used for astronomical calculation, without accepting the physical truth of the heliocentric planetary arrangement.

Mathematical astronomers trained at Wittenberg, notably Erasmus Reinhold (1511–53), Caspar Peucer (1525–1602) and Rheticus, whom we have encountered already, adopted this approach. In passing, we may note that Reinhold was appointed Professor of Higher Mathematics at Wittenberg in 1536, and his astronomical tables, calculated using Copernican techniques, were published in 1551 as the Prutenic Tables, because they were sponsored by the early Lutheran ruler, Albert, Duke of Prussia (1490–1568). Courtly patronage could work alongside academic appointment. Peucer became Professor of Mathematics at Wittenberg in 1554, then also accepted the Chair of Medicine in 1560, a combination we have seen in the career of Gemma Frisius.

Courts, Churches, and Colleges

The use of Tycho's island had been granted by the Danish King Frederick II (1534–88). Later in his career (in 1599) Tycho became Imperial Mathematician to Emperor Rudolph II (1552–1612) in Prague. His experiences indicate the importance of courtly and royal patronage, even though he was himself a nobleman. The mathematical specialisms that might seem most useful to a prince lie perhaps in the arts of war and in surveying, architecture, or water management on a large scale, but astrology in particular brought astronomy into play, with all its mathematical content. The point is reinforced by Kepler succeeding Tycho as Imperial Mathematician in Prague. Another example of Habsburg patronage is the biography of the Italian mathematician Janello Torriani (*c.* 1500–85), who had an extended career in the service of the Spanish monarchs Charles I (Emperor Charles V, 1500–58) and Philip II (1527–98), as court clockmaker, engineer, mathematician, and, through his planetary machines, astronomer and astrologer.

Kepler's service for Rudolph certainly suited his somewhat maverick and obsessive character and the relative freedom it granted to his creativity was exceptionally productive. It is much harder to imagine him reforming the fundamental program for astronomy, had he been tied to the disciplinary assumptions and expectations of a typical university curriculum. The mathematician Jost Bürgi (1552–1632), who did original work on trigonometry and on logarithms, and was a brilliant instrument and clockmaker, was employed by William IV of Hesse-Kassel (1532–92) (and later by his son Moritz, 1572–1632), who had a well-equipped observatory, and then by Rudolph in Prague, where he collaborated with Kepler. An "unlearned" man such as Bürgi (used in the sense of the period to indicate a lack of Latin), however brilliant, would not have found a university post, but his career was sustained through the independent possibilities of courtly patronage.

In Italy, Galileo Galilei (1564–1642) sought to throw off the limitations of his position as Professor of Mathematics at the University of Padua and, like Kepler, wanted to move beyond the restrictions of mathematics as the discipline was currently defined and to engage with the ideas and challenges of natural philosophy. It was in this role that he offered his services to the Medici dukes of Florence, even if it was less than clear what use a natural philosopher would be at court. However, in 1610, Galileo shrewdly used his discovery of four moons of Jupiter, which he named the "Medicean Stars," to secure his ambition.

We can note that the same distinction affected the career in Paris of the mathematician Peter Ramus (1515–72), though in quite a different way. He obtained teaching positions at various colleges of the University of Paris but his anti-Aristotelian stance made this difficult in logic and natural philosophy. Mathematics did not lead him into this kind of conflict. His later career was troubled in many ways, in relation to philosophy and religion, but at each turn, mathematics, where he adopted a practical bias, was a much more manageable option.

Like Galileo, the Venetian Giovanni Benedetti (1530–90) strayed from mathematics into aspects of natural philosophy, though with less determination and persistence. His career also was mainly in ducal service, with some university teaching. He did not attend university himself but did study geometry under Niccolò Tartaglia (1500–57, of whom more later). Though independently wealthy, he accepted the position of court mathematician to Duke Ottavio Farnese (1524–86) at Parma, where his duties included designing sundials, engineering works and astrology. In 1587, he took a similar position with Charles Emmanuel I, Duke of Savoy (1562–1630), while giving lectures at the University of Turin. He published on sundials, geometry, perspective, music, motion and mechanics, and his inclination to introduce issues of natural philosophy was perhaps, like Galileo, related to the disciplinary tolerance of courtly service.

Church patronage also offered possibilities for mathematicians. Antonio da Sangallo the Younger (1484–1546) might seem an unlikely candidate for mathematical advancement, having first set out on an apprenticeship as a carpenter in his native Florence, but after moving to Rome managed to secure an architectural apprenticeship and then to attract commissions. In 1520, he followed such luminaries as Bramante (*c.* 1444–1514) and Raphael (1483–1520) by taking charge of the project of building St. Peter's Basilica, while other commissions included fortification design. Not all papal architecture was ecclesiastical. Was he a mathematician? The geometrical skill required for such challenging classical design work would suffice for a positive answer, but recent work on surviving drawings show him working also on mathematical instruments—a form of sector, long before the designs of Galileo, the astrolabe, and the shape of the paper gores to be applied to a globe (Huppert 2014).

Mathematical expertise in this area could also be found *within* the church. Daniele Barbaro (1514–70) was a scholarly mathematician and humanist, who published a translation of Vitruvius with a commentary; he displays it with evident satisfaction in his portrait by Veronese (1528–88) now in the Rijksmuseum, Amsterdam (he was also painted by Titian, *c.* 1490–1576)—two volumes almost obscuring an armillary sphere, one opened at an illustration of columns, the other at a design for a sundial. Barbaro was a significant churchman as well as a practicing architect, representing his native Venice at the Council of Trent and being secretly appointed cardinal by Pope Pius IV (1499–1565). A major mathematical book on perspective includes an account of the astrolabe, because its projection has the same basis in geometry. Barbaro left an unfinished book on sundials and other mathematical instruments, along with his own instruments for astronomy.

As an example from the seventeenth century, some religious orders, notably the Jesuits, had sufficient strength in mathematics to have members design their churches, and a group of "priest-mathematician-architects" has been identified in the history of architecture (Klaiber 2014). Probably the best known is Orazio Grassi (1583–1654), Professor of Mathematics at the Collegio Romano, who designed several significant Jesuit churches and is familiar to historians of science as Galileo's target in his polemical work, *Il Saggiatore*.

The church also provided mathematical employment to astronomers in the context of calendar regulation and reform. This could extend to clerics with specialist ability, and Copernicus himself maintained his devotion to astronomy in his position as a cathedral canon. The urgent need of reform to the Julian Calendar became a preoccupation in the middle decades of the sixteenth century and a number of mathematical astronomers were consulted. The

Italian Dominican priest, Egnatio Danti (1536–86), was a mathematician with commitments to astronomy, cosmography and architecture, and a mathematical author who designed and constructed instruments, some of which can be admired in museum collections today. His career combined his priestly calling, princely and papal patronage, and university appointments, together with workshop practice and astronomical observation. Danti held chairs of mathematics at Pisa and at Bologna, undertook commissions in cosmography and in architecture from Cosimo I de' Medici (1519–74), tutored the Medici princes, was appointed papal cosmographer and mathematician, and served on the pope's commission for the reform of the calendar. He also worked on large-scale hydraulic, engineering, and surveying projects for his patrons. Danti was an exemplary instance of how to forge an elite mathematical career through a range of activities, opportunities, and appointments.

The chief architect of the Gregorian calendar reform, on the technical side, was the Jesuit mathematician and astronomer, Christopher Clavius (1538–1612), the long-serving Professor of Mathematics at the Jesuit college in Rome. As a teacher and writer of books for mathematical instruction, Clavius had a profound influence, not to mention his championing of mathematics in the curriculum for all the Jesuit colleges. He was active in the design of instruments, especially sundials, while his substantial book on the astrolabe is a powerful example of how an instrument could be the basis of a large and detailed textbook in practical mathematics. Clavius is a further instance of the mathematician as textual editor, through his editions of Euclid's *Elements* and of the *Sphericks* of Theodosius of Tripoli. A portrait of Clavius by Francisco Villamena (*c.* 1565–1624) is an evocative image of the mathematician of the period, thoroughly engaged with his books and instruments, which are given equal importance in the iconography (see Figure 2.2).

The Jesuit order was important for education in advanced mathematics through its network of colleges in Europe and through its missionary work farther afield. In addition to being teachers, Jesuits were often active in practical mathematics, astronomy, and cartography. Perhaps the most famous example of mathematics supporting missionary endeavor comes from a pupil of Clavius, Matteo Ricci (1552–1610) in China, where he lived from 1582 until his death in 1610. One of his notable achievements was a Chinese edition of the first six books of Euclid's *Elements*, translated from the Latin text of Clavius.

Ricci's origins were Italian, and he had cultural links with Portugal, while inheritors of his pioneering mathematical tradition in China illustrate the breadth of the European origins of Jesuit mathematicians. Johann Adam Schall von Bell (1591–1666) was a German, who came to China in 1619 and such was the success of the Jesuits in astronomical prediction that he was appointed president of the Board of Mathematics, which had the important responsibility of promulgating the Chinese calendar. Later in the seventeenth century, the

FIGURE 2.2. Printed portrait of Christopher Clavius, by Francesco Villamena, 1606. Photo by Heritage Art / Heritage Images / Getty Images.

Flemish mathematician and astronomer Ferdinand Verbiest (1623–88) was also involved with calendar reform. He followed Ricci's example by translating the first six books of Euclid's *Elements* into Manchu. In 1673, he built a new observatory in Beijing, though, since he adapted the designs of Tycho Brahe, the instruments were old-fashioned by Western standards.

These men were specialist mathematicians as well as missionaries but a general mathematical competence in the order, thanks to the work of Clavius, had an effect. In 1623, Nicolò Longobardo (1559–1654) was one of two Portuguese makers of a large terrestrial globe, a typical European cosmographical artifact. He had succeeded Ricci as Superior General of the Jesuit mission in China and we have no reason to think that he had more than the standard mathematical expertise of members of the order. His collaborator, Manuel Dias

(1574–1659), was more of an astronomer, and published a textbook in Chinese in 1615, where he mentions the telescope. Their globe is heavily annotated in Chinese with information intended to introduce the concepts and content of Western global geography. It survives in the British Library.

If a college might be an instrument of church patronage, the Collège Royal in Paris, founded in 1530 by François I (1494–1547) with a chair in mathematics, is an example of a royal equivalent. The broad scope of sixteenth-century mathematics gave it a role in a range of institutions that might not seem promising of mathematical employment. The early Italian art academies in Florence, Perugia, Bologna and Rome included formal teaching in geometry and other mathematical sciences. Pietro Antonio Cataldi (1548–1626) was the first lecturer in mathematics at the Accademia del Disegno in Florence (founded by Cosimo I de' Medici in 1563), before moving to the college in Perugia, where two of his lectures were published. He ended his career teaching mathematics at the University of Bologna. Clearly a post at an academy for the arts could play a positive role in a mathematical career. Danti taught in Florence, as did Ostilio Ricci (1540–1603), who had previously been an influential teacher of Galileo in Pisa. There is evidence that the newer and more flexible curricula constructed for the academies meant that theoretical strands were more easily accommodated to an audience that might include the nobility as well as practicing artists. The content of university teaching was more limited. Antonio Santucci (?–1613) combined the interests of astronomer, cosmographer, and instrument designer and maker, and a reference to the two spheres he used to illustrate his lectures in 1595 show that he extended this combination to his mathematical teaching at the academy in Florence. Two spectacularly large and complex armillary spheres can be seen today—one in the Museo Galileo in Florence and the other at El Escorial near Madrid.

As a final example in this section, we can stay in Iberia, where the Portuguese mathematician João Baptista Lavanha (1550–1624) maintained a career almost consistently in royal service—successively for Sebastian I of Portugal (1554–78) and Philip II and Philip III of Spain (1578–1621). While he received a variety of duties and commissions, including cosmography, hydraulics, cartography, and teaching mathematics, it is not surprising that navigation was more prominent in his responsibilities than we have encountered so far, and we find him writing treatises on the subject and making relevant instruments. Navigation was an important aspect of mathematical work and one that appears in a range of contexts for professional practice, so we shall return to it later.

Cities and Corporations

Civil authorities, not confined to royal or ducal courts, had need of mathematicians. Employments offered by cities, provinces or republican states tended to concentrate on the practical: the ornamentation of a dynasty, such as the Medici in Florence, or personal interests in the occult, such as those of Rudolph in Prague, were scarcely relevant, while even astrology was of less significance to a more collective governing body. However, time-telling, calendar regulation, metrology, hydraulics, surveying, architecture, warfare, gauging of barrels, and other forms of measurement and calculation for excise were only the more prominent of many relevant areas of mathematical practice. There was, of course, considerable overlap with the concerns of princely authorities, and churchmen and their councils might have commissions for architecture as well as needing mathematical work regarding calendars and time.

One class of practitioner who served all these masters as occasions arose and whose members have received a fair degree of attention from historians in recent years is the "artist-engineers." Whether this is still the best descriptor for them has been part of this attention but it is easier to list some of the characteristic tasks they undertook, such as designing the structures of large buildings and managing the elevation of materials to shape them, designing siege engines and catapults for offence in warfare, diverting rivers and creating canals, designing mundane machines, such as pile drivers, dredgers, or boring engines, or exotic ones such as submarines or even flying machines. Was this mathematics? Boundaries in the general field of practical mathematics are difficult to draw and to maintain, and many of the artist-engineers had serious mathematical interests by any criteria, which were not allocated an isolated space in their range of work.

We inevitably think of Leonardo da Vinci (1452–1519), self-taught in mathematics but through the study of advanced authors, including Leon Battista Alberti (1404–72) and Piero della Francesca (*c.* 1415–92), alongside, of course, Euclid. His illustrations for *De divina proportione* (On the Divine Proportion) of Luca Pacioli (*c.* 1446–1517) must count as a publication in geometry. There are many instructive instances beyond the most obvious one. A major player in contemporary algebra, Niccolò Tartaglia from Brescia in Northern Italy, was also a leading authority in artillery, writing fundamental works in the mathematization of projectile motion, set within the practice of shooting. Another algebraist, Rafael Bombelli (1526–72) of Bologna, did not have a university education but, tutored by the artist-engineer Pier Francesco Clementi (1526–72), he, too, became a hydraulic engineer, mainly engaged in draining land, largely at the behest of the papal authorities, which sustained his work in algebra and at intervals offered the time he needed to write an influential textbook.

Filippo Brunelleschi (1377–1446), the Florentine architect, engineer and perspectivist, pre-dates our period but should be mentioned for his influential role in this area of mathematical practice (cf. Chapter 6). Trained in mathematics, he was able to take a geometrical approach to problem-solving in architecture and perspective, and his intellectual associates included other mathematicians such as the cosmographer Paolo dal Pozzo Toscanelli (1397–1482) and the humanist and architect Alberti. The character of Brunelleschi's intellectual circle and the techniques he brought to his work evince a strong mathematical aspect to his public persona, which had an important influence on the identity adopted by his successors. While much of Brunelleschi's work arose from ecclesiastical commissions, one of his most influential architectural projects was the Foundling Hospital in Florence, commissioned by the Silk Merchants' Guild, to which Brunelleschi himself belonged. This introduces a further source of practical mathematical employment and patronage, namely companies and guilds in the corporate lives of cities. It is also noteworthy that the famous tradition of the Sienese artist-engineers was born in the context of a republic.

If we imagine we have found at least one pattern to mathematical careers, a counterexample will appear to challenge it. Guidobaldo del Monte (1545–1607) of Urbino looks like another typical artist-engineer, coming from a prominent military family, having experience in the army, publishing on mechanics and on perspective, concerned with mathematical instruments and accepting a position of surveyor of fortifications for Ferdinando I de' Medici (1549–1609). In fact he was a wealthy aristocrat with no need of employment or patronage. Variety again wins the day.

Trained as a weaver, Johann Faulhaber (1580–1635) of Ulm was the social antithesis of Guidobaldo, even if they overlapped in expertise. Faulhaber became city mathematician and surveyor, in 1600 opened a school in Ulm, undertook commissions for fortifications elsewhere and made mathematical instruments, especially for military use. He published in algebra.

Mathematicians might be engaged by civic authorities undertaking major public works, such as water supply, canals, river management, bridges, buildings, and so on. Other commissions were less predictable. Mercator's wall map of Flanders, 1540, was commissioned by a group of merchants in Ghent, as a collective attempt to influence its dedicatee, Emperor Charles V.

Ludolph van Ceulen (1540–1610) owed much to the city councils of, successively, Delft and Leiden, who encouraged his founding of fencing schools where he also taught mathematics. Building on these opportunities, he was later appointed the first Professor of Mathematics at the new Engineering School of the University of Leiden, so becoming a university professor with no university education and no knowledge of Latin, and adding surveying and fortification to his curriculum. We should note also that positions at the Engineering School sustained the careers of successive members of the van Schooten mathematical family: Frans the Elder (1581–1645), pupil and successor to van Ceulen, Frans Junior (1615–1660) and the latter's half-brother Pieter (1634–79).

For a seventeenth-century example, Jonas Moore (1617–79) is an exemplary instance of the metropolitan mathematician (Willmoth 1993). With no university education, he learnt mathematics from books and the encouragement of friends, becoming himself a teacher, tutor, and writer of mathematical textbooks, based in London. He escaped from a routine pattern of work through his appointment as surveyor to a company formed by the Duke of Bedford for draining the Fens; in addition to effectively directing the work, he produced a magnificent large *Mapp of the Great Levell of the Fenns* (1658). Further teaching, publishing, and surveying commissions in London followed, such as a map of the Thames for the Navy Board, and eventually the major appointment of his career, as Surveyor General to the Ordnance. He went on to publish on fortification and artillery. His success allowed him to become a patron himself, in the career of John Flamsteed (1646–1719) and the provision of instruments for the newly founded Royal Observatory at Greenwich. He earned his contemporary nickname, "Mathematical Moore."

In the city, there might be other regulatory bodies and companies whose interests were relevant to mathematics. To return briefly to Nuremberg, the "compass-makers" (in fact, makers of portable sundials) came to form a recognized craft within the centralized legislation that controlled manufacture in the city, and this craft was dominated by a few families (Gouk 1988). The administrative system in Nuremberg was unusual and in the long run their typical formulation of strict rules for such a craft was not well suited to mathematical instruments. Many cities had independent guilds or companies, established by charter from the civic authorities. Guilds for clock- and watchmaking, for example, might claim jurisdiction over the makers of sundials, as they did in seventeenth-century London, and by extension over the work of all mathematical instrument makers (Turner 2008).

The kind of regulation that governed, or sought to govern, instrument making differed between cities. Being relatively new and not easy to define in terms of its materials, techniques, or areas of application, not least because this last aspect was continually under revision, the trade might remain unregulated in certain cities. Elsewhere, for example in Paris, it might come under the protection of the monarch or the church. Finally, it might indeed fall under the regulatory authority of a guild, but even then this could happen in different ways. In London, for example, instrument makers were expected to belong to a guild and to have apprentices bound under its rules, but it did not seem to matter which guild was used. The Clockmakers or the Spectaclemakers seemed appropriate, but many instrument makers belonged to the Grocers, the Joiners or any number of other possibilities.

The strangeness of the instrument makers' case derived from the role of mathematics in underpinning their practice. Most trades could be defined by the materials they used (such as goldsmiths), by characteristic techniques (such as embroiderers or printers) or by the finished product (for example, candle-makers); mathematical instrument makers were linked by the use of geometry and their trade was sustained by the claim that this technical resource empowered expansion into new fields and innovative products. In such circumstances it might seem that the regulatory authority of companies and guilds would be little more than a tiresome obligation in the lives of mathematical instrument makers, but it brought them into a professional context for administrating their role in the city, and into personal contact with other makers and other trades, especially where, as was usual, there was no specific company for mathematical work. They had professional and commercial skills to learn from brother members, should they be merchant tailors, goldsmiths, or grocers. In relation to workshop practice, since their numbers were relatively small and since the usual parameters were not relevant to controlling their methods and materials, instrument makers were generally left alone to innovate as they wished.

However the manufacturing trades were managed in particular locations, the corporate aspect of city life affected the professional activity of this class of mathematical specialists, in regard to training, qualification through apprenticeship, and office holding. Anthony Turner has introduced a distinction between the "mathematical artisan" and the more familiar "mathematical practitioners," but it is often difficult to maintain, as many cases would fall into both. Elias Allen (*c.* 1588–1653), Henry Sutton (*c.* 1624–65) and Walter Hayes (*c.* 1618–before 1696?), makers with contemporary repute in mathematical practice, would be only a few on the considerable list for London, where the distinction would not be helpful.

Other kinds of corporate bodies could be relevant to the professional lives of mathematicians. Trading companies depended on technical expertise in navigation, surveying, cartography, and hydrography. The farther afield their ambitions carried them, the greater their dependence. Their corporate headquarters needed reliable records of their distant interests and grounded, consistent accounts of the routes there and back, with techniques for following them, which often involved sun, moon, and stars. Since the earth was spherical and maps and charts were flat, geometrical cartography was essential, alongside a sound appreciation of how to use these records as mathematical tools. Agents of the company needed skill and knowledge in managing their voyages safely and efficiently. This required astronomy as well as geometry and routines for calculation, so mathematicians were needed to devise instructions and to give lessons.

Both Portugal and Spain established institutions in the sixteenth century with responsibilities for technical maritime matters. In Lisbon, the Armazém da Guiné e Indias (Guinea and Indies Warehouse) was in charge of instruments, cartography, and the training of pilots, and needed staff with relevant expertise. Separately, a position of Cosmographer (Chief Royal Cosmographer by 1547) was given to the mathematician Pedro Nunes (1502–78), who pursued a career between these responsibilities and appointments at different universities, while publishing in algebra, cosmography, and navigation, and improving instrument design. In Spain, the Casa de Contratación (House of Trade) was established in Seville in 1503 and among its responsibilities were the training of pilots and the preparation and updating of maps and charts. Here, the pilot-major in the mid-sixteenth century was Pedro de Medina (1493–1567), whose textbook *Arte de navegar* (Art of Navigation) was widely republished and translated.

In the mid-century, England was beginning to form ambitions for overseas activities in emulation of Portugal and Spain and some appreciated the need for a similar cultivation of practical mathematics. The Muscovy Company was an early trading enterprise, chartered in London in 1555 for trade with Russia, and the mathematician Robert Recorde (*c.* 1512–58) became an adviser on technical matters such as navigation. He dedicated his algebra text, *The Whetstone of Witte* (1557), to the Company. Dee took up this role and advised the Company on navigation.

The Dutch East India Company (Vereenigde Oostindische Compagnie, VOC), created by an amalgamation of companies in 1602, could sustain its extensive enterprise only by maintaining a leading position in navigation and cartography. The ships' masters were well trained and the company produced their own charts. Willem Blaeu (1571–1638), who had studied with Tycho Brahe and established a map and globe business in Amsterdam, was appointed company cartographer in 1633. He was succeeded in this position by his son Joan (1596–1673) in 1648, as the Blaeu family continued their central position in the flourishing Dutch trade in map and globe-making through much of the seventeenth century. The mathematician, Johannes Hudde (1628–1704), who had studied with Frans van Schooten in Leiden, became governor of the company in 1679.

The (English) East India Company (EIC), incorporated in 1600, had similar technical needs through the seventeenth century. These stimulated publication and private teaching in navigation, but cartography was more advanced in the Netherlands and there the English depended in this period on the Dutch. The first governor of the EIC, Sir Thomas Smythe (*c.* 1558–1625), had been central to a group of London merchants founding a mathematical lectureship, occupied by Thomas Hood (1556–1620) (Charlton 2007), and the Company later paid Edward Wright (1561–1615) a salary for lecturing on navigation.

Official boards, commissioners, and offices of state had other work for mathematicians. Ordinance, shipbuilding, excise, and public works all provided either posts or temporary commissions. "Events" also, of course, played their part. We shall see in the following section that the Fire of London profoundly altered the careers of two prominent mathematicians.

PRINCIPLES, TOOLS, AND TRENDS

Mathematical Sciences and Arts

We have considered the character of mathematical practice through a selection of individual careers. A complementary narrative can operate at the level of disciplinary development more generally, not least because there was a widespread adoption in the period of a program for the expansion of mathematical science into a range of arts that would themselves become mathematical in nature. Why should that be a desirable end? For mathematicians, it would expand their reach and relevance. It could also enhance social status: it has been argued plausibly that the coming of the canon and, in response, of geometrical fortifications raised the status of mathematicians as military engineers and as tutors to the aristocracy (Biagioli 1989). For the practitioner, they were promised enhanced reliability, accuracy, certainty, and status. The price for them was the need to learn new methods and perhaps to buy lessons, books and instruments. Not everyone was convinced, but sufficient numbers of practitioners accepted the proffered opportunities to begin a process of profound change.

In any case, compromise and even bluff or dissimulation were often possible. A good example comes from the ambitious surveying textbook, *The Surveyor*, by the mathematician Aaron Rathborne (*c.* 1570–after 1622), published in 1616. An elaborate title page (see Figure 2.3) depicts a mathematical surveyor making a show of using the most challenging novelty in the contemporary effort to mathematize surveying practice, an altazimuth theodolite.

In the text, however, Rathborne does not even describe the instrument, because it is so challenging, and admits that he uses the much simpler plane table, even though such a practice is mocked on the same title page. Instruments themselves illustrate this mathematical sham, for while they might have arcs for measuring both horizontal and vertical angles, it might also be possible readily to remove the altitude arc before use, leaving a much more manageable instrument for horizontal angles only.

If mathematicians had set their sights on turning surveying into a mathematical art, not least through the promised advantages of triangulation, the same was true of navigation, gunnery, fortification, gauging and architecture, while a number of arts were already coming under their sway, such as perspective drawing, cartography, gnomonics, and so on. This trend widened the scope for variety and flexibility in individual careers. Christopher Wren (1632–1723) was a mathematician who held the Savilian Chair of Astronomy at Oxford, while accepting occasional commissions in architecture and maintaining interests in sundials, perspective, and surveying. Official appointments following the Fire of London in 1666 altered the course of his life and reputation, but no-one questioned whether he was suited to such a prominent architectural responsibility—it seemed that a mathematician was just what was needed. Robert Hooke (1635–1703), despite his multifaceted identity, was enough of a practical mathematician to move in exactly the same direction.

If a dominant register of mathematical work in the period was expressed through the mathematical arts, there was disciplinary space for activity that did not have particular practical or material relevance, and might in contemporary

FIGURE 2.3. Title-page from Aaron Rathborne, *The Surveyor*, 1616. Photo by Science & Society Picture Library / SSPL / Getty Images.

English be called "speculative." There was no problem about working and publishing in both: alongside his treatise *Pantometria* (1571), on the subject of surveying and peppered with designs for instruments, Thomas Digges (*c.* 1546–95), protégé of Dee, published a geometrical account of the Platonic solids. Digges saw military service in the Netherlands and published an original account of military mathematics, his *Stratioticos* (1579). The celebrated Italian algebraist Girolamo Cardano (1501–76), who had a fractured academic career and supplemented his income through medicine and gambling, must be thought of as a speculative mathematician but contributed to astrology and practical mechanics. A significant reputation could be achieved through speculative mathematics: François Viète (1540–1603) was never employed as a mathematician and on that basis denied that he was one, presumably in a professional sense, but successfully pursued mathematics while earning a living as a private tutor, a lawyer, and a royal councilor.

Alongside shared methods and ambitions, there was sharing of philosophical assumptions. Something similar to the "Wittenberg interpretation" we have encountered in astronomy appears in other areas of practice, giving what can be called an "instrumentalist" role to mathematics. In cartography, for example, maps and charts of large areas or even of the whole earth could be projected differently for different purposes, resulting in different shapes for the map and for its contents. Since cartography was a mathematical discipline, there was no reason to ask which of these projections was the correct one. They did not aim at physical veracity, but at a combination of technical or procedural correctness (it would be misleading to use the modern notion of rigor) and aptitude to particular uses.

The word "theoric" (*theorica*) was used to refer to these kinds of "instruments," literal as well as metaphorical. Most theorics took a mathematical form: they might be a construction, a diagram, a map drawn by projection, or a physical instrument on brass or paper. Their use yielded results without an assumption that the means of generating results reflected a reality in the material world. The best known examples would be the planetary theorics of Ptolemaic astronomy, which yielded the apparent positions of the planets in the sky for given dates, times, and geographical locations. Copernican mathematical astronomy also could be treated as a theoric. The mathematician's association with this instrumental approach became part of the accepted character of the discipline.

Books and Instruments

Books and instruments were prominent and effective tools for both furthering mathematical practice and promoting professional careers for both author and reader. To say that books could be an important element in building a mathematical vocation might, with hindsight, seem an obvious truism, but it is worth considering why mathematics was an early beneficiary of the invention of printing. A reliable version of a text could be distributed widely, could be available at any time for individual study, by purchase or through a library or an obliging acquaintance. This text could be followed at a speed to suit the reader, could be poured over, with passages reread as necessary. It was particularly helpful that geometrical diagrams were easily supplied by including woodcuts; these were, of course, a very significant aid to individual study. All of this is obvious now but only because mathematics was so readily adapted to print. We have seen that much mathematical practice had involved active engagement—a matter of carrying out procedures based on mathematics and learning to do so through practice. The role of books was an important element in the gradual process of making individual study the main characteristic of learning mathematics. The idea that books were a substitute for other forms of mathematical learning is reflected in the presentation of a number of early elementary texts as dialogues between a master and a scholar.

Examples where the publication of books was important for a mathematical reputation are too numerous to mention. Cardano's reputation benefitted greatly by his successful publications. The works of Nunes on navigation had an international reach through translation. Recorde had a career as an academic, a physician and an agent of the Crown in mining and coinage, but is credited with enormous influence in the foundation of mathematical practice in England through his popular, vernacular textbooks.

Instruments were closely associated with mathematics in this period, so closely that we should not refer to "scientific" instruments, as that is a much broader category: the only class of instrument for much of the period covered by the present chapter was the mathematical. The class was, however, much broader than the modern "mathematical instrument," by which we generally refer to instruments for drawing or calculation, or perhaps some more exotic devices such as integrators or planimeters. A "mathematical instrument" in the sixteenth or seventeenth century could refer to any instrument used in a mathematical art, such as quadrants, sextants, or astrolabes in astronomy, theodolites or plane tables in surveying, cross-staves or nocturnals in navigation, sundials or horary quadrants in horology. We have seen already that these instruments, unlike the later telescopes, microscopes, air-pumps and so on, were not for investigating the natural world, but for use in the mathematical arts and for the convenient solution of problems there. Their role, in other words, was consistent with that of mathematics itself, as distinct from natural philosophy.

Books and instruments were closely related. Many mathematicians were concerned with the design of instruments and many more published books treating them—either accounts of existing ones or descriptions of the virtues and properties of new ones. Stöffler's influential book on the "construction and use" of the astrolabe, mentioned above, was exactly what it said—the first half took the reader carefully through the process of making an astrolabe, the second half through worked examples of the many ways to use it and the many problems it could solve. This was an important precedent: it became common to adopt the "construction and use" pattern in books on instruments.

The relationship could be closer still. We have mentioned the paper instruments, some constructed as "volvelles" with moveable parts, in cosmographical books by Apian and by Gemma Frisius. Other mathematical authors did the same; it was clearly an attractive feature for book buyers. As early as 1476, the final page of the *Kalendarium* (Calendar) of Regiomontanus had a printed universal rectilinear altitude sundial (the Regiomontanus dial), complete with an articulated brass arm riveted to the page, for setting the suspension point of the plumbline to the appropriate latitude and date.

The Regiomontanus instrument was, of course, a woodcut, but when engraved copperplates came to be used in printed books, much more ambitious and accurate instruments printed on paper could be bound into volumes or sold as accompanying sheets. Once a set of projected lines had been laid out for an astrolabe or horary quadrant, engraving on brass was the principal skill of the mathematical instrument maker, and this could be adapted to working in reverse on copper for printing in a rolling press. Sutton, for example, would carry this process to remarkable ends and produce some of the finest paper instruments in the mid-seventeenth century, including his own design of horary quadrant. While he was the engraver, he insisted on being recognized also as the inventor of the projection and designer of the instrument, and therefore a mathematician.

The importance of instruments to the practice of mathematics conveyed importance also to the instrument maker. To take a single example, the London "shop" (which in English referred equally to a space for making or for selling) of Elias Allen became a rendezvous for practical mathematicians and a place for the exchange of information and ideas—another example of a mathematical *locus*. It was used as a postal address and the preface to Moore's *Arithmetick* textbook of 1650 was dated: "From my Chamber at Mr. Elias Allen, his house over against St. Clements Church in the Strand. 30th. of October, 1649." It was also vital to the careers of prominent mathematicians who created innovative instrument designs to find a maker who would step outside his routine production. In this way, Allen became essential to the careers of William Oughtred (1575–1660) and Edmund Gunter (1581–1626), both of whom designed instruments for calculation by logarithms.

Societies, Journals, Correspondence and Contests

The seventeenth century saw the beginning of societies formed for the cultivation and promotion of experimental philosophy. This has long been a topic of great interest to historians of science, but how did it affect mathematicians? Their discipline sat at a distance from natural philosophy but the relationship was beginning to change.

The story is usually begun with the *Accademia dei Lincei* (Academy of the Lynxes), founded and supported by the aristocrat Federico Cesi (1585–1630), and active in Rome between 1603 and 1630. There were a few mathematicians among its number, notably Luca Valerio (1552–1618), who taught mathematics at the university, and famously Galileo, who had proposed Valerio. The

Accademia del Cimento (Academy of Experiment), founded in 1657 in Florence, was also supported by patronage, in this case from the Medici family. Activities were focused on experimental philosophy but two leading members, Giovanni Alfonso Borelli (1608–79) and Vincenzo Viviani (1622–1703), were mathematicians. We have already encountered the Italian academies of the arts as teaching establishments but they also had aspects of learned societies (Skaarup 2015). The Florentine *Accademia del Disegno* (Academy of the Arts of Drawing) numbered Galileo among the members, as well as Evangelista Torricelli (1608–47) and Viviani. Academy membership entailed more than teaching: with Medici patronage, an appointment was also an index of status.

If the *Accademia* would not usually be found in the standard history of science, the account of mathematics emerging in this chapter suggests that another type of institution might play a part. Cabinets of curiosity, especially those containing natural wonders, have been a focus of interest in recent years, but there were also mathematical cabinets, which were fashionable slightly earlier. If experimental academies were sites of patronage and display, mathematical cabinets shared aspects of their character and function, though located on the other side of the division between art and nature. A Florentine example would be the *Guardaroba* (now *Sala delle Carte Geografiche*, Wardrobe, or Hall of Geographical Maps) of the Palazzo Vecchio, designed as an immersive cosmography, which held the Medici collection of mathematical instruments. Another example would be the precursor collection of what is now the Dresden *Mathematisch-Physikalischer Salon* (Mathematical-Physical Cabinet) of the Electors of Saxony, in its formative period in the sixteenth and seventeenth centuries.

The Royal Society of London was founded in 1660, following a lecture at Gresham College (see below) by a mathematician, Wren, and appointed a mathematician, William Brouncker (1620–84), as its first president. The society first called itself a "College for the Promoting of Physico-Mathematical Experimental Learning." All of this seemed to bode well for mathematics in the society but, although present, it was not a strong element in the program. Perhaps it was not readily adapted to the society's conventions of communal discussion. Only a handful of fellows elected in the first decade will have been admitted only for their mathematics, such as John Pell (1611–85), William Neile (1637–70), Nicolaus Mercator (*c.* 1620–87), and John Collins (1625–83). On the other hand, a great many elected fellows in this period showed little interest in science of any kind.

The Académie des Sciences in Paris, founded in 1666, did give stronger formal recognition to mathematics, with sections for three mathematical sciences—geometry, mechanics, and astronomy—matching three physical sciences (chemistry, botany and anatomy). Academicians were awarded pensions (essentially paid a salary). Mathematical members included Pierre de Carcavi (*c.* 1600–84), Bernard Frenicle de Bessy (*c.* 1604–74), Gilles de Roberval

(1602–75) and from the Netherlands Christiaan Huygens (1629–95). The Academy was generous to mathematicians and particularly mathematical astronomers, whose importance was underlined by the building of a magnificent observatory, completed in 1672. By contrast, royal patronage was sought in London but granted in name only.

At the opposite end of the spectrum of formality, we know of several relatively casual mathematical clubs or societies meeting in London in the seventeenth century. In each case, the record is a casual remark or a piece of ephemera. One met at the Bull's Head tavern in Cheapside and is known only by a printed slip of paper from 1681, addressed to Pell, inviting him to dine with them, on condition that he pay the bearer two shillings and sixpence and "bring this Ticket with you" (Beeley 2019). Without patrons or publications, societies of this sort will mostly have passed unrecorded.

The *Philosophical Transactions* did not carry much mathematics in its early years. Issues in the 1660s do have some mathematical papers, such Brouncker's quadrature of the hyperbola, and Wren's demonstration of the generation of the hyperboloid of revolution by a straight line, so mathematics was present but not strongly so. In contributing to a mathematical life, the *Acta Eruditorum*, published in Leipzig from 1682, was more significant than either the *Philosophical Transactions* or the *Journal des sçavans*, but this came largely after our period. The early volumes of the annual *Connaissance des temps*, published in Paris from 1679, did not restrict its content to the formal ephemerides.

Exchange of correspondence was a common feature of intellectual life in the period, used for the interchange of views and even the announcement of results. Letters could be shared and copied on, and this practice was transferred to societies. A good proportion of Royal Society meetings was devoted to the reading of letters received by the secretary. A leading participant in this republic of letters was the mathematician Marin Mersenne (1588–1648), who lived in Paris as a monk in the Order of the Minims and, in addition to his religious calling, was active in facilitating communication across the range of mathematics and natural philosophy. He was famous for his extensive correspondence, by which he kept others informed as well as himself, and he organized meetings for discussion and exchange of reports, which became known as the Academia Parisiensis. Mathematicians who benefitted from being in his circle included René Descartes (1596–1650), Gilles de Roberval, Blaise Pascal (1623–62), and Girard Desargues (1591–1661). Other significant figures who each carried on a large correspondence included Johannes Hevelius (1611–87), Pierre de Fermat (1601–65), John Wallis (1616–1703) and Christiaan Huygens. Letters were an important tool for coherence and identity in the scattered and, by everyday standards, esoteric world of mathematicians.

Finally, we must attend to contests. Tartaglia's career as a mathematician, for example, was profoundly influenced by challenges and contests around his ability to solve cubic equations. Mathematicians were not always cooperative: they could be challenging and competitive, but it can be argued that even this created identity and, after a fashion, community. It was possible to engage in such activity only if you belonged to the same discipline.

In 1593, the mathematician Adriaan van Roomen (1561–1615) of Louvain, recently appointed Professor of Medicine at Würzburg, proposed a 45th-degree equation as a challenge to the whole mathematical world. It was solved by Viète, as a favor to his patron Henry IV (1553–1610), in defense of French mathematics, and he proposed in turn the problem of drawing a circle to touch three given circles, known as the problem of Apollonius, to which van Roomen published a solution in 1596. A seventeenth-century example with nationalistic overtones was the so-called "Jean de Montfert" problem, originating in France under this pseudonym as a challenge to English mathematicians. Wren had his solution printed as a broadside, setting the French a reciprocal challenge to solve "Kepler's problem," which related to the astronomer's so-called second law. Wren had a geometrical solution, later published by Wallis, who wrote of the French mathematicians, "none of them have yet (that I hear of) returned any Solution" (Hall 1965).

Contests could also be intensely personal, while carried on in public. In 1633, the Dutch mathematical teacher, tutor and author, Jan Stampioen (1610–90), set Descartes a specific challenge involving a quartic equation, and then rejected Descartes's solution as incomplete. Stampioen later presented public challenges under an alias and then published solutions under his own name. One solution offered by a surveyor called Waessenaer to a problem in ballistics was rejected by Stampioen and the subsequent adjudication itself became mired in argument involving Descartes, who is thought to have been implicated in the Waessenaer solution in the first place. This type of activity could both divide and consolidate: national and personal reputations might be threatened but such challenges relied on a wider assumption of disciplinary identity and mutuality.

Extending Northwards

Our narrative has strayed into the seventeenth century from time to time, in our choice of examples to illustrate the diversity of mathematical careers, but do any overall trends emerge in the latter part of our timeframe?

One feature is a northwards extension of mathematical practice, with the rise of ambitions in the Netherlands and in Britain for expanding political and trading activity that would require mathematical expertise in navigation, surveying and the arts of war. Commercial expansion would in turn stimulate mathematical practice at home, in business, and excise and in civil, military, and naval architecture. We have already mentioned both East India Companies—Dutch and English—which became rich and powerful organizations with international reach. Both countries (the English following and imitating the Dutch) were engaged in schemes for hydraulic management and drainage.

For the turn of the century we might cite the Flemish mathematician and military engineer Simon Stevin (1548–1620) as a leading figure in practical mathematics and point to his military service and publications across the familiar range of mechanics, architecture, fortification, navigation and so on, as well as in more "speculative" topics. We might set him beside his contemporary in England, Thomas Harriot (*c.* 1560–1621), who had practical experience of the sea and, although he did not publish in mathematics, taught navigation and left extensive manuscript work also in astronomy, optics, instrumentation, mechanics, cartography, surveying, and speculative mathematics. Both studied at university—Leiden and Oxford respectively. Both were commissioned and employed by leading political figures, Stevin by Prince Maurits (1567–1625), Stadtholder of the United Provinces, Harriot by Sir Walter Ralegh (1554–1618) and then by Henry Percy, Earl of Northumberland (1564–1632).

Both Leiden and Oxford saw the founding of new professorships dedicated to mathematics in the seventeenth century. These created options in mathematical careers but, more than that, they gave recognition to the discipline, the prospect of a figurehead role for a successful incumbent and the possibility of a tradition in mathematical practice associated with an institution. The first mathematical professorships in England were not at a university, however, but at Gresham College, founded in London in 1597, with chairs in geometry and in astronomy and statutes that included topics in practical mathematics. Henry Briggs (1561–1631), the first Professor of Geometry worked especially on the logarithms of John Napier (1550–1617), which were of great use to his calculations in astronomy and the provision of navigational tables. Briggs was appointed to the new Savilian Chair of Geometry in Oxford in 1619, thus neatly demonstrating the career possibilities these positions had created and their potential for forming a platform for a significant mathematician. The fact that the posthumous printing in England of Briggs's logarithmic tables was overseen in 1633 by Henry Gellibrand (1597–1637), Gresham Professor of Astronomy, illustrates how a tradition of work can be sustained through the continuing existence of an institution. Wright, mentioned already, was an associate of Briggs and an important member of the growing English tradition in the mathematics of navigation, who, like Harriot, had had seaborne experience in the Atlantic.

Wright's academic associations had been in Cambridge, where a mathematics chair, the Lucasian, followed Oxford in 1663, with Isaac Barrow (1630–77) appointed to it; Barrow had been Professor of Greek at Cambridge and also of Geometry at Gresham College. Meanwhile the Gresham astronomy professors had included several mathematicians, Edmund Gunter, Samuel Foster (*c.* 1600–52), and Christopher Wren, and the Savilian Chair of Astronomy had accommodated mathematicians Seth Ward (1617–89) and (again) Christopher Wren. Interests in mathematical instruments and other aspects of practical mathematics are strongly represented in this group, which was the case also for Barrow's successor as Gresham Professor of Geometry, Robert Hooke. Wallis occupied the Savilian chair in geometry for over fifty years, from his highly political appointment in 1649, while with the kind of opportunist diversity we have come to expect, his career was advanced through church appointments (he was an ordained minister), by becoming keeper of the University archives, and in a time of conflict by his activities in cryptography.

Mathematics professors at Leiden included Willebrord Snell (1580–1626), succeeding his father Rudolph (1546–1613) in 1613, and Albert Girard (1595–1632), a French Protestant and religious refugee. When appointed in 1629, he was already Professor of Arabic, and as well as being an algebraist, he spent time in the Dutch army as an engineer, working especially in fortification. Jacobus Golius (1596–1667) also simultaneously held chairs in mathematics and Arabic at Leiden.

An influential mathematical career was still possible outside the academic, collegiate, or civic spheres. Oughtred had a substantial following among the younger English mathematicians. He built a livelihood on a modest career in the church, combined with the patronage of the Earl of Arundel, giving private lessons, writing books and, as we have seen, an association with Allen, who made and sold the instruments he designed.

CONCLUSION

Within a general agreement over the character of mathematical arts and sciences, we have encountered a great variety of ways to profess mathematical practice and live out a career in the discipline. There was an unruly element to mathematics as a profession, ironic perhaps in a subject that prized reason and rigor. We have found a great range and variety of routes for being a mathematician and no standard career path to follow (Biagioli 1989). A number of *loci* fostered a mathematical identity, but individual choices, sequences, or combinations among them were diverse. In the latter part of our period, this unruly character was beginning to decline, as institutional influences grew, from universities, societies, and more centralized administrations.

This was an exciting and challenging time to have been involved with mathematical practice, a time of development and opportunity. The basic idea that mathematical techniques could be common to a range of practical activities and could be adapted to the reform of a growing number of useful arts was a sound one and had a fair measure of success. As we have seen, this enabled mathematicians to move between different disciplines and to accept employment and commissions in a variety of contexts, from universities and colleges to princely courts and church and civic authorities.

Success bred confidence—assurance that an underlying mathematical science could be developed in its own right and would prove valuable for its own sake and for its relevance to the increasing range of mathematical arts. This assurance underpinned the vitality with which mathematicians began to approach that fundamental disciplinary and professional divide between their legitimate practice and natural philosophy. Newton's *Principia* has been taken as the endpoint of this volume and its full title, "Mathematical Principles of Natural Philosophy," indicates how far the program of mathematical penetration into other disciplines had become normalized by 1687. No one accused Isaac Newton (1643–1727) of a confusion of categories in the way Maestlin had challenged Kepler. Quite the opposite: he was challenged instead to provide a physical explanation of gravitation, without which his mathematical account seemed arbitrary and mysterious.

CHAPTER THREE

Inventing Mathematics

Betwixt Tradition and Innovation

SÉBASTIEN MARONNE

Our historical journey begins with the first printed version of Euclid's *Elements* based on the Latin translation by a medieval scholar, Campanus of Novara (*c.* 1220–96). This text was published in 1482 and initiated the long and erudite tradition of printed editions of works in ancient Greek geometry. Our survey ends two centuries later, in the 1680s, with the influential works of two giants torn between tradition and innovation: Gottfried Wilhelm Leibniz (1646–1716) and Isaac Newton (1643–1727). Infinitesimal calculus, as it developed in the first decades of the eighteenth century (cf. Chapter 3 in Volume 4), originated from the contributions of both men. In 1684, Leibniz published a brief and abstruse article entitled *Nova methodus pro maximis et minimis, itemque tangentibus* (A New Method for Maxima and Minima, and for Tangents) in the scholarly journal *Acta Eruditorum*. His presentation used the symbolism and the techniques of modern algebraic analysis. Three years later, Newton published *Philosophiae naturalis principia mathematica* (Mathematical Principles of Natural Philosophy), which took its shape from ancient Greek geometry, though supplemented by infinitesimal considerations. Both of these approaches—involving algebraic as well as infinitesimal practices—had been previously introduced and developed throughout the seventeenth century.

These objective boundaries thus point to a long-lasting and continuous process reflecting an interplay of traditions dating from the classical mathematics of the ancient Greek geometers to the innovative mathematics of the Moderns. By Moderns, we simply mean those mathematicians of the early modern period who called themselves Moderns, as did the protagonists in the famous literary quarrel between Ancients and Moderns at the end of the seventeenth century. In mathematics, this quarrel was complex and colored the entire period. While René Descartes (1596–1650) identified himself as an outspoken modern, Leibniz emphasized how his methods went beyond Descartes and even refuted some of his claims. Newton, on the other hand, whose *Principia* was taken as the last word by many, was fond of the Ancients and attempted to refute Descartes's claims for the superiority of the "new analysis" over Greek methods.

In order to describe this process and to provide an overview of mathematical invention in the early modern period, we have chosen to follow several paths up to the exemplary epilogue offered by Newton's *Principia* and the birth of Leibnizian calculus. The dialectical tensions between Ancients and Moderns, which provided a strong incentive for inventing mathematics, will be illustrated by several examples. The method of analysis, both as an older and newer method of invention, will be discussed, on the one hand, alongside topics such as the classification of problems and curves, on the other. During the period under consideration, the focus was on solving problems, and analysis seemed to be an efficient method in this respect. Among the various ways of writing mathematics, symbolic language occupied an important and growing place. Symbolic algebra was used to solve arithmetical as well as geometrical problems. Finally, we will describe the new methods of indivisibles and infinitesimals, which were devised for handling problems involving tangents and quadratures.

THE DIALECTICS BETWEEN ANCIENTS AND MODERNS

Throughout the early modern period, the mathematical community acquired access to the great works of ancient Greek geometers (Euclid, Apollonius, and Archimedes), but also to the fragments compiled by Pappus of Alexandria (*c.* 290–350), through various critical editions, Latin translations, and prolix mathematical commentaries. Two key protagonists in this movement were the German Jesuit Christopher Clavius (1538–1612) and the Italian mathematician Federico Commandino (1509–75). The former provided an influential Latin edition of Euclid's *Elements* (1574), whereas the latter edited Apollonius's *Conics* (1566) (see Figure 3.1) and Pappus's *Mathematical Collection* (1588), providing each with a detailed commentary.

While the Euclidean tradition had little influence on mathematical invention in the seventeenth century (even if it played a normative role in the writing of mathematics), the works of Apollonius and Archimedes had a major impact. Apollonius's work on conic sections (*Conica*) and Archimedes's mathematical and mechanical works (*Measurement of a Circle*, *The Quadrature of the*

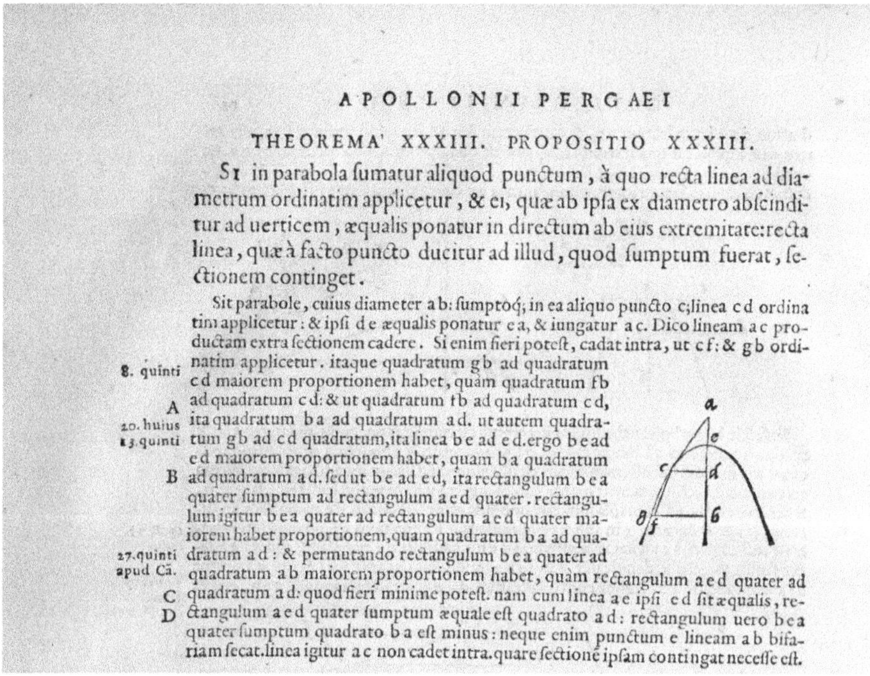

APOLLONII PERGAEI

THEOREMA' XXXIII. PROPOSITIO XXXIII.

Si in parabola ſumatur aliquod punctum, à quo recta linea ad diametrum ordinatim applicetur, & ei, quæ ab ipſa ex diametro abſcinditur ad uerticem, æqualis ponatur in directum ab eius extremitate: recta linea, quæ à facto puncto ducitur ad illud, quod ſumptum fuerat, ſectionem continget.

Sit parabole, cuius diameter a b: ſumptoq; in ea aliquo puncto c; linea c d ordinatim applicetur: & ipſi d e æqualis ponatur e a, & iungatur a c. Dico lineam a c productam extra ſectionem cadere. Si enim fieri poteſt, cadat intra, ut c f: & g b ordi-

8. quinti natim applicetur. itaque quadratum g b ad quadratum c d maiorem proportionem habet, quàm quadratum f b

A ad quadratum c d: & ut quadratum f b ad quadratum c d,

20. huius ita quadratum b a ad quadratum a d. ut autem quadra-

13. quinti tum g b ad c d quadratum, ita linea b e ad e d. ergo b e ad e d maiorem proportionem habet, quam b a quadratum

B ad quadratum a d. ſed ut b e ad e d, ita rectangulum b e a quater ſumptum ad rectangulum a e d quater. rectangulum igitur b e a quater ad rectangulum a e d quater maiorem habet proportionem, quam quadratum b a ad qua-

27. quinti dratum a d: & permutando rectangulum b e a quater ad

apud Cã. quadratum a b maiorem proportionem habet, quàm rectangulum a e d quater ad

C quadratum a d: quod fieri minime poteſt. nam cum linea a e ipſi e d ſit æqualis, re-

D ctangulum a e d quater ſumptum æquale eſt quadrato a d: rectangulum uero b e a quater ſumptum quadrato b a eſt minus: neque enim punctum e lineam a b bifariam ſecat. linea igitur a c non cadet intra. quare ſectionẽ ipſam contingat neceſſe eſt.

FIGURE 3.1. Apollonii Pergaei Conicorvm Libri Qvattvor. Bononiae: Benatius, 1566. Courtesy of Max Planck Institute for the History of Science, Berlin.

Parabola, *On the Equilibrium of Planes*) stimulated new research on geometrical curves. The theory of conic sections led to further studies of tangency problems, whereas the methods of Archimedes promoted interest in quadrature problems. Tangency problems were about finding the tangent straight line to a geometrical curve at a given point. Quadrature problems asked for the area of a curvilinear figure by comparing it to that of a given rectilinear figure, for instance a given square or a triangle. Mathematicians of the early modern period were also tempted to restore lost treatises, especially those described by Pappus in his *Mathematical Collection*. Thus, in the 1630s, Pierre de Fermat (1601–65) gave a modern version of Apollonius's *Plane loci*, a work devoted to geometrical problems whose solutions were either straight lines or circles.

In the reception and interpretation of ancient works, the concern for faithfulness sometimes conflicted with new ways of writing mathematics that increasingly relied on symbolism. For instance, Isaac Barrow (1630–1677), the Lucasian Professor in Cambridge, did not hesitate to introduce symbolic calculations in his edition of Apollonius's *Conics* of 1675 in order to provide shorter algebraic versions of geometrical demonstrations. In any case, the classics often provided a stimulating propaedeutic for mathematical invention, not only with regard to the selection of problems but also with respect to methods for solving them as well.

It should be noted that this interplay between commentary and problem-solving, tradition and innovation, happens to be quite rare in the history of mathematics. Another example, however, can be found a few centuries earlier in Arabic mathematics, which was important for early modern mathematics not only on a conceptual level but also by way of textual transmissions. In 1629, the Dutch orientalist and mathematician Jacob Golius (1596–1667) brought a number of Arabic mathematical manuscripts back to Leiden from his travels in the Orient. These included the *Algebra* of Omar Khayyām (1048–1131) and the Arabic translation of Apollonius's *Conics* made by the Banū Mūsā during the ninth century. The latter included Books V–VII, which were lost from the original Greek version. As will be shown below, there was a close relationship between early modern mathematics and the classical Greek tradition. Both shared strong interests in methods of invention, the classification of problems and curves, and the conceptual foundations of theories, despite the apparent novelty of their respective lineaments.

The only form of formal exposition recognized in Greek geometry was synthesis. A synthetic demonstration proceeded by means of the axiomatic method, i.e. by relying either on axioms or on previously proved propositions until it produced the statement to be proven. Note that this statement had to be known beforehand. As a systematic presentation of knowledge acquired by other means, synthesis was valued for its rigor and its certainty. In contrast, analysis was a method of invention and proceeded in the reverse order. It assumed either the existence of a sought object or the validity of the theorem to be proved and drew inferences until it reached an already known object or statement. Then the synthesis could be produced by arguing in the reverse direction (Bos 2001: 95–118).

Greek geometrical analysis was summarized by Pappus in a rather obscure passage that appears at the beginning of Book VII of his *Mathematical Collection*. Pappus's *Mathematical Collection* was significant not only as a rich and eclectic catalogue of geometrical problems, accompanied by various solutions proposed by Greek geometers using curves, instruments, or other auxiliary constructions. It was also the main source for partial accounts of numerous lost works of Greek geometry, on which the restitutions mentioned earlier were grounded. Especially intriguing for moderns was the fact that Pappus referred to a "treasury of analysis" that was comprised mainly of lost works, whereas the extant texts of Greek geometry were all written using synthesis. This contrast led the mathematicians of the early modern period, many of whom were not content with synthetic demonstrations, to imagine that the ancients had an analytical method of invention for solving geometrical problems, but that they had kept this method a secret.

Thus, the algebraic analysis initiated by François Viète (1540–1603) in his *In Artem analyticem Isagoge* (Introduction to the Analytic Art, 1591) and elaborated by René Descartes in his *Géométrie* (Geometry, 1637) was not seen by its authors as entirely new. Instead, it was conceived by them and others as a kind of restored classical analysis, at least on a rhetorical level, and overlooking that fact that these two approaches were in some ways quite different. The algebraic analysis of a geometrical problem reduced it to an equation in which the symbols stood for both known and unknown lengths of lines. The careful examination of diagrams in classical analysis, which Descartes considered to be "greatly tiring for the imagination," was replaced by the clever handling of equations. One should recall here that his *Discourse on Method* was a seminal work for both philosophy *and* science. Indeed, the *Geometry* was the third essay he appended to his text in order to show how his general method could be applied "to seek truth in the sciences." Moreover, while the ancients gave pre-eminence to synthesis over analysis for the sake of rigor, early modern geometers like Descartes claimed that analysis was superior to synthesis, not only as a heuristic method but also on a demonstrative level.

Nonetheless, classical geometrical analysis did not vanish with the emergence of algebraic analysis; on the contrary, it continued to be used. Thus, the Croatian mathematician Marinus Ghetaldi (1568–1626), although a disciple of Viète, took this traditional stance in his treatise *De resolutione et compositione mathematica* (On Mathematical Resolution and Composition, 1630). Ghetaldi thought that classical analysis (*resolution*) surpassed algebraic analysis in the resolution of certain geometrical problems, and so he also kept providing the synthesis (*composition*). Toward the end of our period, Isaac Newton wrote his Lucasian Lectures on Algebra, which were published against his will by William Whiston (1667–1752) in *Arithmetica Universalis* (Universal Arithmetics) in 1707. This text appeared to the reader to be a somewhat paradoxical handbook devoted to the application of the (Cartesian) method of algebraic analysis to the resolution of geometrical problems. Indeed, in it, Newton criticized the exaggerated importance Descartes attached to the algebraic criterion in the construction of geometrical problems, and he even claimed that, in some cases, Apollonian geometry should be preferred to the Cartesian (on these matters, see Guicciardini 2009). A few years later, Newton would decide to write his *Principia* (1687) in accordance with the classical synthetic geometric style.

Another aspect of mathematical invention at the crossroads of tradition and innovation, beyond the debate on methods, arose from the classification of geometrical problems. In Greek geometry, mathematicians distinguished between problems and theorems. When solving a problem, one constructed the object to be found, whereas the demonstration of a theorem proved a property attributed to a given object. In Book I of Euclid's *Elements*, for instance,

Proposition 46 was a problem: it asked the reader "to construct a square on a given straight line." Proposition 47 then stated the so-called Pythagorean theorem by considering the squares that could be constructed on the sides of a right-angle triangle by virtue of Proposition 46.

A famous classification of geometrical problems was based on the curves used for their solution; this can be found, again, in a passage from Book IV of Pappus's *Mathematical Collection*. According to Pappus, *plane problems* were those that required circles and straight lines for their solution, *solid problems* were those solved by means of conics, and *linear problems* required other curves such as the conchoid, cissoid, spiral curves, or the quadratrix. More important still, it would be a "sin" (*peccatum* in Commandino's Latin translation) to use an inappropriate curve in the solution of a geometrical problem, for example, a conic or a linear curve when the problem was planar. Mathematicians of the early modern period devoted much attention to this classification, including Pappus's stringent precept, and they proposed their own classifications. They also dealt with geometrical problems, such as the three classical Greek problems, namely duplication of the cube, trisection of the angle, and quadrature of the circle. These problems had been solved in Book IV of Pappus's *Mathematical Collection* by using constructions other than by ruler and compass. The constructions of Ancients and Moderns employed curves, especially conics, mechanical instruments, or special methods, like the *neusis* (intercalation), which consisted in fitting a line segment between two straight lines from a given point. These problems thus required methods beyond those of the Euclidean canon, i.e. constructions by ruler and compass. The demonstrations of the impossibility of solving the three classical Greek problems by ruler and compass would only be given much later in the nineteenth century.

Among the innovative mathematicians of the seventeenth century, Descartes played a prominent role. His classification of geometrical problems, which went beyond the one described by Pappus, was based on the degree of the equation obtained at the end of the algebraic analysis of the problem, and he paired this with a classification of curves. In a famous passage that inaugurates Book 2 of his *Geometry*, Descartes proposed a new classification for "genres" of "acceptable curves in geometry," based on the degree of their equation. According to Descartes, the main criterion for "geometrical curves" was that they be expressible by an algebraic equation. On the other hand, they needed to be generated geometrically by one or several continuously linked motions; one could then construct all their points by intersecting the curve with an appropriate circle (see below). Finally, Descartes excluded curves like the quadratrix or spirals from geometry. These transcendental curves, which met none of the above criteria, would later be put back into play by Leibniz and his followers because of their growing importance in differential calculus with its many applications to physical phenomena.

These two related problems—the classification of problems *and* curves—are thus quite exemplary for the dialectical relationship between ancient and early modern mathematics. They, indeed, bear witness to an old and unbroken tradition, which began with Greek geometers and was developed further by Arabic mathematicians. It then reached a climax with the classification of cubics given by Isaac Newton in his *Enumeratio linearum tertii ordinis* (Enumeration of Lines of the Third Order). This work was composed already in the late 1670s, but only published in 1704 as an appendix to Newton's *Opticks*. During this era, classification criteria, both geometric and algebraic, continued to be placed in balance and weighed against each other within ongoing methodological discussions.

Further illustration of the fruitful exchanges at the intersection of tradition and innovation arose from discussions concerning the problem of the dimension of geometrical quantities. Within classical Greek geometry, the objects were situated either in a plane or in the three-dimensional space of our physical intuition. Geometers then studied their corresponding magnitudes, namely lengths, areas, and volumes. Moreover, only *homogeneous* magnitudes of the same dimension could be compared or combined. In modernizing geometry, Viète and Descartes went beyond those classical limits, though in different ways. In his *Introduction to the Analytic Art* (1591), Viète gave a generalization of the older theory of magnitudes. He thus introduced "specious logistics" as a system of abstract quantities of arbitrarily large dimension that satisfied the usual operations of addition and multiplication. In this system, however, the homogeneity constraint still applied, whereas the dimension of the product of quantities was the sum of their dimensions. Thus, the product of two line segments was a rectangle, but one could only form sums of quantities if they had the same dimension.

Descartes circumvented this homogeneity restriction by introducing an arbitrary unit. In the very first lines of his *Geometry*, Descartes "related arithmetic to geometry" by defining multiplication, division, and extraction of square roots for line segments, once a line segment had been chosen as a unit. This allowed him to get rid of the homogeneity constraint since, unlike Viète, the multiplication of two line segments produced another line segment. Nevertheless, one should not imagine that Descartes's algebra of line segments was adopted at once. Many mathematicians who read the *Geometry* continued to use Viète's algebra (as did Fermat, for instance). Even those who used Cartesian algebraic analysis often ignored the conceptual foundations of the algebra of line segments, and Descartes himself often considered homogeneous equations. By doing this, one could more easily check algebraic calculations and provide constructions independent of the choice of unit.

Finally, the theory of ratios and proportions, as presented in Book V of Euclid's *Elements* and applied to similar plane figures in Book VI, provided both a technical basis for mathematical practice as well as a privileged theme for epistemological discussions. Indeed, several basic theorems of classical geometry that played a key role in mathematical practice were stated in terms of proportions. In natural philosophy, Galileo Galilei (1564–1642) used geometrical theorems to study the uniformly accelerated motion of free-falling bodies (cf. Chapter 5). He published these findings in his influential *Discorsi e dimostrazioni matematiche intorno a due nuove scienze* (Discourses and Mathematical Demonstrations Relating to Two New Sciences, 1638). Difficulties arose, however, when mathematicians of the early modern period dealt with the complex definition of equal ratios due to Eudoxus and taken up by Euclid in Book V of the *Elements*. This definition was often misunderstood and its interpretation gave birth to some heated debates.

SOLVING PROBLEMS

In early modern mathematics, greater emphasis was placed on the solution of problems rather than on the demonstration of theorems. When confronted with geometrical problems, such as tangency or quadrature problems, mathematicians often asked the following questions: What does it mean to solve a geometrical problem? Which methods are legitimate? When can we assert that a geometrical entity is known or given? Which constructions are the most legitimate in geometry, beyond those using ruler and compass? No demonstration or technical argument could be advanced to answer such questions definitively. This meant that "quality" constraints regarding the resolution of problems had to be discussed on a methodological and epistemological level, even if the arguments provided were mathematically rooted. This led, in effect, to a *diversity of mathematical cultures and practices*, despite the common ground provided by classical geometry.

Problem-solving underwent a major transformation with Descartes's *Geometry*, and this change was by no means merely technical. There is no doubt that Descartes's new way of using algebraic analysis had a tremendous impact, in particular by drawing attention to the importance of a theory of curves per se. A key ingredient involved what we today call a "Cartesian reference frame" (though with some differences, see below). Descartes introduced this in his *Geometry* to solve a *locus* problem transmitted by Pappus, which he reduced to a second-degree equation in two unknowns by means of algebraic analysis (cf. Figure 3.2). This enabled him to *demonstrate* that the solution was a certain type of conic (parabola, hyperbola, or ellipse) depending on the given data. Only a few decades later, mathematicians would no longer regard the equation of a curve as posing a problem indicating a task to be performed (namely, constructing the curve) but rather as a solution, which provided a representation of the curve. This shift in understanding reflects an important change in habituation associated with Descartes's work, now considered to be a landmark in the history of mathematics.

Yet, even if historians have rightly emphasized the role of Descartes's *Geometry* in this conceptual process, it should be noted that his masterpiece provoked controversies after its publication.

Another characteristic change brought about by Descartes's *Geometry* concerns what the historian of mathematics Henk Bos has called the redefinition

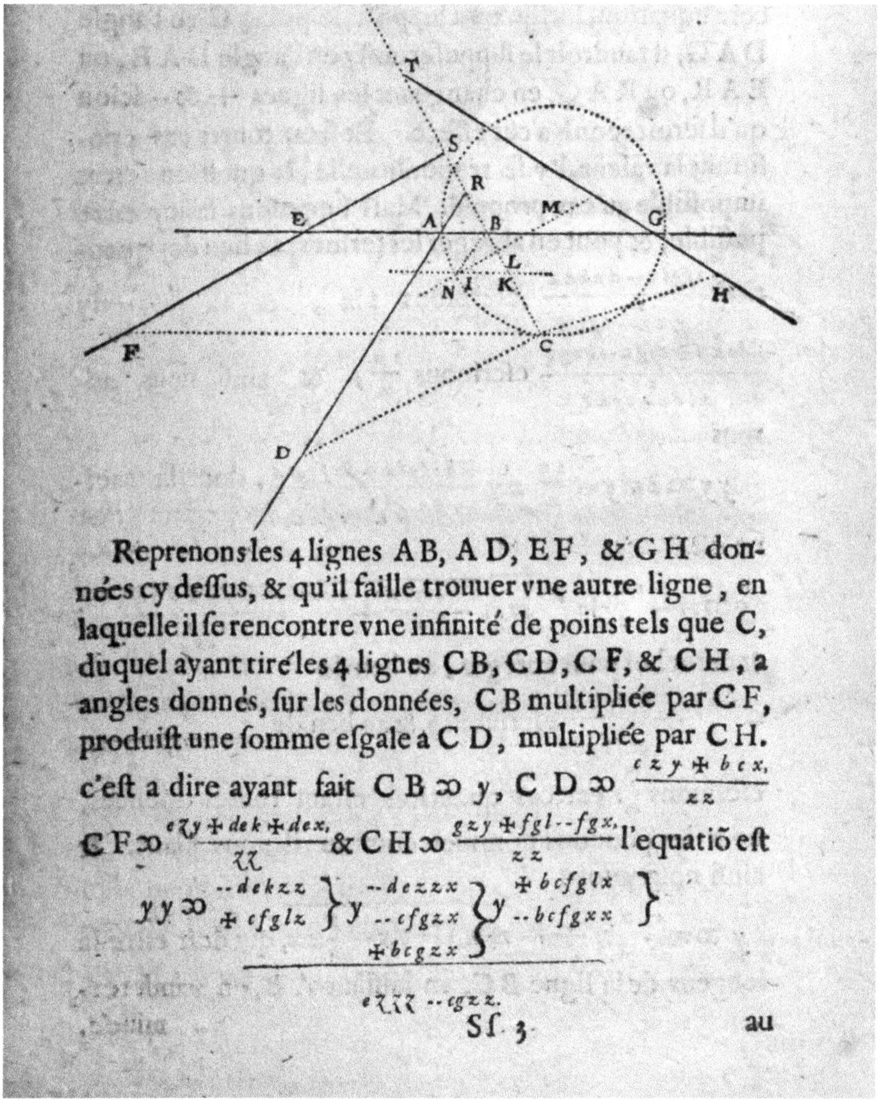

Reprenons les 4 lignes AB, AD, EF, & GH données cy dessus, & qu'il faille trouuer vne autre ligne, en laquelle il se rencontre vne infinité de poins tels que C, duquel ayant tiré les 4 lignes CB, CD, CF, & CH, a angles donnés, sur les données, CB multipliée par CF, produist une somme esgale a CD, multipliée par CH. c'est a dire ayant fait $CB \propto y$, $CD \propto \frac{czy + bcx}{zz}$, $CF \propto \frac{ezy + dek + dex}{zz}$ & $CH \propto \frac{gzy + fgl - fgx}{zz}$ l'equatiõ est

$$yy \propto \frac{\left.\begin{matrix} -dekzz \\ +cfglz \end{matrix}\right\} y \left.\begin{matrix} -dezzx \\ -cfgzx \\ +bcgzx \end{matrix}\right\} y \left.\begin{matrix} +bcfglx \\ -bcfgxx \end{matrix}\right\}}{ezzz - cgzz.}$$

Sf. 3 au

FIGURE 3.2. *AB* and *BC* are the abscissa and the ordinate of point C. Descartes's solution of Pappus's locus problem in *Discours de la méthode pour bien conduire sa raison, & chercher la verité dans les sciences. Plus La dioptrique, les météores et la géométrie, qui sont des essais de cete méthode* (1637: 325). Bayerische Staatsbibliothek München: Rar. 1849.

of geometrical exactness (see Bos 2001), namely the way in which Descartes refined the methodological and epistemological criteria that governed the resolution of geometrical problems. Indeed, once a geometrical problem had been put in an algebraic equation, one then needed to find the roots of that equation in order to determine the unknown straight lines which were required to construct the solution of the problem. This was exactly Descartes's claim in the first sentence of the *Geometry*: "Any problem in geometry can easily be reduced to such terms that a knowledge of the lengths of certain straight lines is sufficient for its construction" (trans. Smith and Latham [1637] 1954: 2). But what did it mean to *know* the "lengths of certain straight lines"? Here, to *know* did not mean to provide an algebraic formula involving the extraction of roots for quadratic and cubic equations, but rather to provide a *construction* of straight lines of such length as to solve the original geometrical problem. Sometimes, it could also mean to advance an argument that would convince the reader that such a construction was possible.

To this purpose, Descartes constructed the roots of such equations by intersecting a circle and a certain curve composed according to the degree of the equation. For instance, he constructed solutions of third- and fourth-degree equations by intersecting a circle with a parabola. And so on, so said Descartes! However, the possibility of extending this approach to higher-degree equations was anything but self-evident. The pre-eminence Descartes gave to algebraic equations over geometrical constructions in his classification also came under criticism. Fermat discussed this on a technical level in his *Dissertatio Tripartita* (Tripartite Dissertation) written *c.* 1643 and posthumously published in the *Varia Opera Mathematica* and later by Guillaume de L'Hôpital (1661–1704). To take one example, Descartes contended that an equation of degree 7 should be constructed, by means of a circle and a quartic (a curve of order 4), whereas his critics thought one should use cubics (curves of order 3), which are *simpler* curves according to the algebraic criterion.

On the other hand, a striking consequence of Descartes's algebraic approach was its *generality*. Indeed, algebra allowed Descartes to construct a generic equation whose coefficients were arbitrarily given data; the equation then arose from the algebraic analysis of the geometrical problem under study. This method thus enabled him to provide a general solution, which applied to *any* geometrical problem leading to an equation of the same degree. However, there was a downside to this generality: since Descartes operated with generic coefficients, he did not provide the *actual* construction of some geometrical problem, but merely showed the *possibility* of such. So, what happened "in a concrete case" if one wanted to perform the construction proposed by Descartes's general method and to display the solution? This might prove to be rather difficult or perhaps even practically impossible to achieve. This was the case, for instance, with a famous problem in *Tangencies*, a lost work by

Apollonius that was described by Pappus. The problem requires the construction of a circle tangent to three given circles, which Descartes solved in his correspondence of 1643 with the German Princess Elisabeth of Bohemia (1618–80). He wished to prove that this was a plane problem, so Descartes contented himself with showing that the algebraic analysis led to a quadratic equation, which meant the problem could be solved by straight lines and circles, as he had demonstrated in his *Geometry*. On the other hand, he provided neither the equation nor the highly complicated construction.

Thus, according to Descartes, it was enough to exhibit the equation, i.e., to show the *nature* of a problem, thanks to the classification noted above. For him, it was by no means necessary to perform the construction and the synthetic demonstration, since doing so unfruitfully duplicated and hid the algebraic analysis. Once again, such a methodological precept did not find general consensus in the early modern period. Indeed, Newton was one of Descartes's staunchest opponents, as we have already seen. Although he perfectly mastered the Cartesian algebraic analysis on the *technical* level and had moreover cleverly developed it further, for instance in his classification of cubic curves, Newton harshly criticized the Cartesian program on the *methodological* level.

WRITING MATHEMATICS

During the seventeenth century, mathematics circulated not only through printed books but also via manuscripts and often in letters. It was only during the second half of the century that scholarly journals were founded and began to play a role in the dissemination of mathematical ideas. These periodicals arose alongside the new scientific academies and societies, like the *Philosophical Transactions of the Royal Society*, which was published from 1665 onward, and the *Histoire et Mémoires de l'Académie des Sciences*, first issued in 1699. Another influential journal, the *Acta Eruditorum*, was founded in 1682 in Leipzig with the support of Leibniz, who published several important articles in it.

Mathematical innovations could thus be found in various kinds of texts, both printed *and* manuscript, as throughout the period scientific correspondence played a crucial role in the dissemination and discussion of mathematical innovations. Marin Mersenne (1588–1645), a Minim monk who lived in the order's monastery in Paris, acted as a privileged go-between among the scholars of France and the whole of Europe, thanks to his correspondence network. Excerpts from letters and appended writings were copied and communicated, thereby influencing the ways in which mathematics was written. As a result, mathematical knowledge was packaged in epistolary rhetoric that reflected both the private and public status of such exchanges. On the one hand, mathematicians were often reluctant to present their inventions in a clear and open way in order to avoid being robbed. A famous case of this from 1638 involved Fermat's method in *De maximis et minimis* (On Maxima and Minima), which circulated among Mersenne's correspondents. This set off a sharp controversy with Descartes concerning the general problem of constructing tangents to curves. On the other hand, the social status of the practitioners clearly influenced the nature of these mathematical exchanges. In particular, mathematical technicalities (especially the use of symbolism) were not always appreciated.

Attention was often paid to the "*honnête homme*," those learned men who were interested in science and mathematics, but easily repelled by overly technical presentations or more exhaustive discussions that did not allow them to engage with the matter at hand. The secretary of the Prince of Orange and father of Christiaan Huygens (1629–95), the famous mathematician and physician, Constantin Huygens (1596–1687) was a typical representative of such figures. He was acquainted with Descartes and enjoyed reading the physical essays appended to the *Discourse on Method,* but he requested the assistance of a private teacher, the mathematician Frans van Schooten the Younger (1615–60), to aid his understanding of the *Geometry*.

Nowadays, we can hardly imagine reading mathematical literature that avoided using appropriate symbolism and formulas. During the early modern period, by contrast, mathematical symbolism fluctuated greatly and often it was nearly absent in texts. The equations that appear in the *Great Art* of Girolamo Cardano (1501–76), which was published in 1545 and dealt with solutions of cubic equations, can be deciphered only with great difficulty today (cf. Figure 3.3). Going back further still, one finds no symbolism at all in Arabic algebra. In al-Khwārizmī's *Kitāb al-Jabr wa-al-muquābala* (Compendium on Calculating by Completion and Reduction), written at the very beginning of the ninth century in Baghdad, equations and algorithms were written in prose. Nonetheless, the numerical problems in the *Arithmetica* of Diophantus of Alexandria (*c.* 201–85) were solved using a proto-symbolic language based on abbreviated designations. These observations reveal that the close relationship we take for granted nowadays between symbolism and algebra, literal calculations, and solutions of equations, only arose after around 1600.

Nonetheless, the various symbolisms that developed during the early modern period in different countries and mathematical schools throughout Europe—English, French, German, Italian, etc. (Cajori [1928/9] 1993)—already converged by the beginning of the seventeenth century. In fact, the notations introduced by Viète and Descartes essentially correspond to those still used today. How can we characterize this style of algebra and discriminate between it and earlier symbolisms? One relevant criterion is provided by the manner in which unknowns were designated. Before Viète, the Italian and German algebraists did not use a specific symbol to denote an unknown quantity. Even if it is quite problematic to speak of *the* inventor of a specific notation for

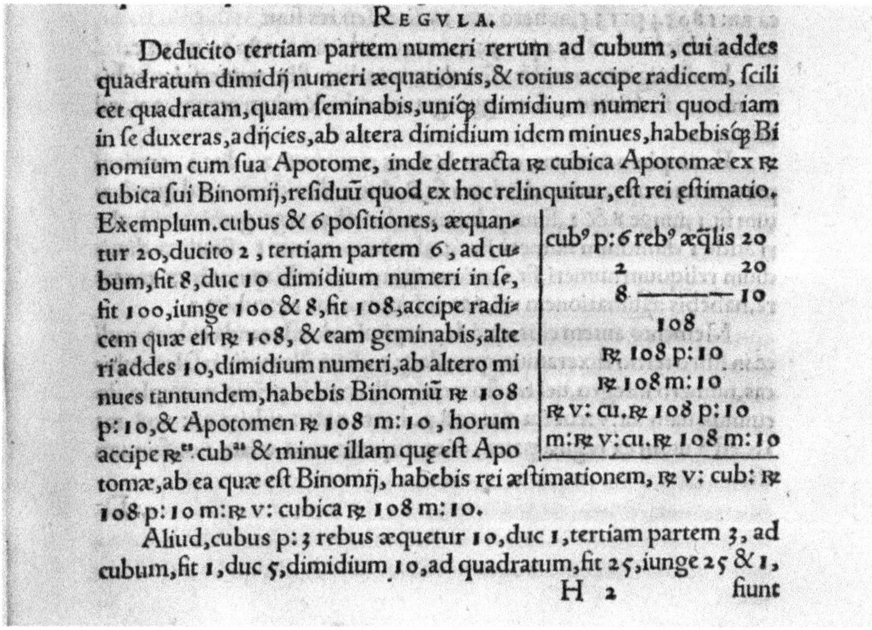

REGVLA.

Deducito tertiam partem numeri rerum ad cubum, cui addes quadratum dimidij numeri æquationis, & totius accipe radicem, scilicet quadratam, quam seminabis, uni[que] dimidium numeri quod iam in se duxeras, adijcies, ab altera dimidium idem minues, habebis[que] Binomium cum sua Apotome, inde detracta ℞ cubica Apotomæ ex ℞ cubica sui Binomij, residuũ quod ex hoc relinquitur, est rei æstimatio.

Exemplum. cubus & 6 positiones, æquantur 20, ducito 2, tertiam partem 6, ad cubum, fit 8, duc 10 dimidium numeri in se, fit 100, iunge 100 & 8, fit 108, accipe radicem quæ est ℞ 108, & eam geminabis, alteri addes 10, dimidium numeri, ab altero minues tantundem, habebis Binomiũ ℞ 108 p: 10, & Apotomen ℞ 108 m: 10, horum accipe ℞ cub & minue illam quæ est Apotomæ, ab ea quæ est Binomij, habebis rei æstimationem, ℞ v: cub: ℞ 108 p: 10 m: ℞ v: cubica ℞ 108 m: 10.

cub⁹ p: 6 reb⁹ æq̃lis 20
2 20
8 —— 10
108
℞ 108 p: 10
℞ 108 m: 10
℞ v: cu. ℞ 108 p: 10
m: ℞ v: cu. ℞ 108 m: 10

Aliud, cubus p: 3 rebus æquetur 10, duc 1, tertiam partem 3, ad cubum, fit 1, duc 5, dimidium 10, ad quadratum, fit 25, iunge 25 & 1,

H 2 fiunt

FIGURE 3.3. The rule for solving the equation $x^3 + 6x = 20$ in Cardano, *Great Art* (1545), Chap. XI, p. 30 *recto*. The solution given is $\sqrt[3]{\sqrt{108}+10} - \sqrt[3]{\sqrt{108}-10}$. Bayerische Staatsbibliothek München: 2 Math, p. 8.

unknowns, Viète is generally considered to hold the honor of having originated such a system, in particular because of the influence of his symbolism. Moreover, Viète's new algebra used vowels "A", "E", . . . to designate unknowns and consonants "B", "C", . . . for the given numbers. In his *Geometry*, Descartes reversed this designation by using the letters at the end of the alphabet "*z*", "*y*", "*x*", . . . to stand for variable quantities, whereas the letters at the beginning "a", "b", "c", . . . denoted given quantities; he also introduced the exponential notation for powers of quantities (cf. the equation in Figure 3.2).

When considering manuscripts or mathematical letters, one can often observe that the mathematicians of the early modern period could easily switch from one symbolic system to another. Moreover, even during the second half of the seventeenth century, one encounters research works *without any* symbolism. A remarkable example comes from the *Lettres de A. Dettonville* (the pseudonym of Blaise Pascal, 1623–62), published in 1659. This tract was devoted to intricate quadrature and cubature problems connected with the cycloid, and yet it was absolutely devoid of symbolic calculations. This suggests why it is necessary to nuance the claim that mathematical innovation necessarily depends on the development of a sophisticated symbolic notation.

ALGEBRA BETWEEN ARITHMETIC AND GEOMETRY

In his treatise on algebra, al-Khwārizmī reduced the general problem of the resolution of quadratic equations to the solving of three species of quadratic equations, each written in prose without any symbolism. The solution of these equations followed from an arithmetical algorithm, which was then justified by a geometrical argument. The very same structure can be found in the works of the sixteenth-century Italian mathematicians—Scipione Del Ferro (1465–1526), Girolamo Cardano, Niccolò Tartaglia (1500–57), and Lodovico Ferrari (1522–65)—who derived solutions of cubic and quartic equations. In Cardano's *Ars Magna* (Great Art, 1545), the resolution of cubic equations was grounded in a classification into thirteen species. For each, Cardano gave a "rule" expressed by an algorithm, followed by a geometrical justification based on an astute arrangement of cubes and parallelepipeds in analogy with the squares and rectangles used to justify the solutions of quadratic equations. The idea behind this was essentially the same in both cases, namely to complete an unknown cube (square) with parallelepipeds (rectangles) in order to obtain a known cube (square). Nevertheless, this new algebra introduced a rudimentary symbolism (cf. Figure 3.3). It also produced new quantities, which *necessarily* resulted from the solutions given by the rule. This led not only to negative quantities but also to so-called "imaginary" magnitudes, i.e. those which we now call non-real complex numbers. Indeed, among the problems tackled by Cardano, one was called the "irreducible case." This involved square roots of negative numbers in the formula for its solution, even though the solution itself was not imaginary, i.e., was a real number, to use modern terminology.

Another influential tradition originated in the editions, commentaries, and translation of the *Arithmetica*, the classic written by Diophantus of Alexandria (cf. Corry 2015: ch. 4). The first six books of the *Arithmetica* were transmitted in Greek to Western mathematicians via Byzantine scholars. In 1968, an Arabic translation by Qusta ibn Luqa (820–912) of four additional books was discovered in Meshed, Iran. The problems in the *Arithmetica* involve finding integers or rational numbers satisfying a certain condition that can be interpreted as an integral polynomial equation, or a "Diophantine equation." A number of editors and commentators promoted this tradition, such as the German classical philologist Wilhelm Xylander (1532–76), who published the first printed Latin translation of the *Arithmetica* in 1575. Another was the French mathematician Claude Gaspard Bachet de Méziriac (1581–1638), who published an influential commentary in his Latin edition of 1621. At the same time, the rediscovery of Diophantus exerted considerable influence on invention by providing challenging problems and new ideas. Before these editions appeared, Rafaele Bombelli (1526–72) benefited from reading a Greek manuscript of the *Arithmetica* in the Vatican Library in preparation for writing his groundbreaking *Algebra* (1572). Later, the Dutch engineer and mathematician Simon Stevin (1548–1620) incorporated a translation of the first four books of Diophantus in his own *Arithmétique* (1585). One could also cite Viète, but by far the most famous innovative reading is due to Fermat, the inventor of the method of infinite descent. This new method based on a proof by contradiction enables one to show that a given "Diophantine equation" has no solutions. Among Fermat's several marginal annotations in his copy of Bachet's edition was the one known to posterity as "Fermat's last theorem." It was only proved more than 350 years later, in the 1990s, by Andrew Wiles. Fermat stated it was impossible to separate any power of a positive integer greater than the second into two like powers of positive integers, *viz.* that the equation $x^n + y^n = z^n$ has no positive integral solutions for $n > 2$.

Beyond number theory, applications of algebra to geometry and physics were especially salient features of mathematical innovation during the seventeenth century and, even more so, throughout the eighteenth. The two profusely annotated Latin editions of Descartes's *Geometry*, supervised by his disciple Frans van Schooten, greatly contributed to the diffusion of Cartesian geometry and its techniques among European mathematicians. Among its innovations, one finds the familiar reference system of coordinates with one axis for the abscissae with its origin and a second axis for the direction of the ordinate. There was no ordinate axis in Descartes's original *Geometry*, but this feature soon became standard and can be found, for example, in Newton's *Enumeration of Lines of the Third Order*. Descartes's *Geometry* differed in another important respect from our modern analytic geometry, since its coordinate systems were intrinsic to a given curve and not extrinsic.

The method of indeterminate coefficients was another important innovation in Descartes's *Geometry* that contributed to the future of mathematical research. On the technical level, it played a key role not only in the theory of algebraic curves but also in the theory of series. The method of indeterminate coefficients was used in comparing two equations. If, for example, one obtained a first equation from algebraic analysis of a problem under study then its coefficients were given, whereas the coefficients of a second equivalent equation were typically *indeterminate*. The identification of the coefficients of these two equations led to a system of linear equations which—once solved in the relevant unknowns—provided the solution to the problem. For instance, when computing normals to an algebraic curve, Descartes identified the coefficients of an equation $P(x) = 0$ derived from the curve equation with those of an equation containing a *double* root for $x = e$, which meant one could write it $(x - e)^2Q(x) = 0$, where $Q(x)$ was a polynomial with indeterminate coefficients. Such an algebraic identification was grounded in a geometrical statement, namely that a certain circle will be tangent to the curve at the point of abscissa

e (and its radius thus normal to it at that point) if and only if it meets the curve in a *double* point.

Although negative quantities took on growing importance in algebraic practice, both in solving equations and in the theory of geometrical curves, their nature and the justification for calculations with them remained problematic. For these very reasons, they were often used only reluctantly or even omitted from consideration, despite the simplifications derived from their (correct) handling. Thomas Harriot (1560–1621) thus neglected negative roots of equations in his *Artis Analyticae Praxis* (Practice of Analytic Art, 1631). Such solutions were designated as "fictitious" in Cardano's *Great Art* or as "false" in Descartes's *Geometry*. Negative coordinates were gradually introduced during the latter half of the seventeenth century, but neither Descartes nor Fermat considered them, unlike Newton, who used negative coordinates in his *Enumeration of Lines of the Third Order*. But what did it mean for a geometrical magnitude to be less than nothing? For those who held to the elementary classical Euclidean theorem, stating that if $a: b = c:d$, then $a > b$ implies $c > d$, a paradox arose at once when negative quantities were introduced. For from $1 : -1 = -1 : 1$, one would derive that $1 > -1$ implies $-1 > 1$! The problem of establishing firm foundations for negative quantities was addressed at length toward the end of the seventeenth century in a controversy between Antoine Arnauld (1612–85) and Jean Prestet (1648–90) (cf. Mancosu 1996: ch. 3). This took place during a time when negative quantities were already incorporated in algebraic practice, independent of such foundational discussions.

FROM INDIVISIBLES TO INFINITESIMALS

The consideration of infinitely small or large quantities played a crucial role in the mathematical innovations of the early modern era. A noteworthy example was Girard Desargues's (1591–1661) *Brouillon Project d'une atteinte aux evenemens des rencontres du Cone avec un Plan* (Rough Draft of Attaining the Outcome of Intersecting a Cone with a Plane, 1639). Desargues, a geometer, architect and engineer from Lyon, introduced points at infinity that went "beyond our understanding" together with a straight line at infinity composed of such infinitely distant points. Parallel straight lines were then considered to meet at some point at infinity, which put them on a par with any other pair of straight lines, since any two always intersected at one point in this geometry. This allowed Desargues to gather together several cases of a geometrical proposition and prove them with a single demonstration. This aim for generality was typical of the early modern period and was embodied in practices as different as those employed by Descartes and Desargues. Indeed, the latter used no algebraic symbolism, but instead he introduced abundant terminology borrowed from botany. For instance, before introducing cross-ratios and defining the involution of six points, Desargues wrote that three pairs of points *B,H; C,G; D,F* lying on the same straight line (*AH*), the "trunk," form a "tree" with respect to a common point *A*, the "stump," when *AB.AH* = *AC.AG* = *AD.AF*. Suffice it to say that none of this made Desargues's condensed opus reader-friendly for his contemporaries. Nonetheless, Pascal acknowledged in his *Essay pour les coniques* (Essay on Conics, 1640) that he owed his discoveries to Desargues's work, and he tried to imitate his methods as far as possible. Another contemporary, the Parisian mathematician Jean de Beaugrand (*c.* 1595–1640), who edited Viète's works in the early 1630s, engaged in a stormy controversy with Desargues. Beaugrand claimed that Apollonius's *Conics* enabled one to solve all the problems tackled by Desargues in his *Rough Draft*.

The geometry of indivisibles invented by Bonaventura Cavalieri (1598–1647) offers yet another example of the dialectical relationship between ancient and early modern mathematics. In his treatise entitled *Geometria indivisibilibus continuorum nova quadam ratione promota* (Geometry, Advanced in a New Way by the Indivisibles of the Continua, 1635), Cavalieri used sums of lines to determine the areas of surfaces. He called these lines "indivisibles" with respect to the surface area. Cavalieri handled problems involving cubatures and centers of gravity in a similar way by using sums of areas. The main principle behind the method of indivisibles consisted of comparing geometric figures by considering the sizes of their corresponding indivisibles. On the one hand, Cavalieri departed from Euclidean orthodoxy by neglecting the homogeneity constraint. On the other, however, he relied on the method of exhaustion of the ancients to ground the method of indivisibles. The method of exhaustion, which was used by Euclid and Archimedes, was a rigorous method of demonstration, but not of invention. It allowed one to solve quadrature problems by means of a double proof by contradiction, without using any infinitesimal considerations. For instance, Euclid proved (*Elements* XII.2) that the ratio between the areas of two circles was neither greater nor smaller than the ratio between the squares on their diameters by using inscribed polygons in the circles (cf. Introduction to Volume 1). As Pascal wrote in a foreword to his *Letters of Dettonville* (1659: 10): "everything that is proved by the true rules of the indivisibles may also be rigorously proved in the manner of the ancients," so that "one of these methods differs from the other only in the manner of speaking."

The new method of indivisibles nonetheless gave rise to objections concerning the rigor of its demonstrations and related calculations. The Jesuit Father André Tacquet (1612–60) asserted in his work *Cylindrica et Annularia* (Cylinders and Annuli, 1651) that an incautious use of indivisibles led to manifestly false demonstrations and that "demonstrations established by the heterogeneous [i.e., the indivisibles with respect to the figure they form] did not meet assent, unless they were reduced to the homogeneous [with respect to the figure], which was possible in most cases" (Book 1, Part 1, Prop. XII, Scholium, p. 24). This adaptation of the method of the indivisibles was proposed by a number of authors, such as Gilles Personne de Roberval (1602–75) in his *Traité des Indivisibles* (Treatise of Indivisibles, posthumously published in 1693 in a collection of his mathematical works by the French Academy of Sciences), Evangelista Torricelli (1608–47) in his *Opera Geometrica* (Geometrical Works, 1644), and Pascal in the *Letters of A. Dettonville* (1659) who replaced indivisible lines with infinitesimal rectangles. More subtly, paradoxes of the infinite lent mystery to the use of the method, which further intrigued mathematicians. A famous example was Torricelli's solid figure resembling an infinite trumpet, an object that had an infinite surface area but a finite volume (Mancosu 1996: ch. 5).

Another impressive example of the use of the method of indivisibles was Torricelli's demonstration of the equivalence of two classical problems: the rectification and the quadrature of the circle. This equivalence had been established by Archimedes in the first proposition of his *Measurement of the Circle* by means of the method of exhaustion. Archimedes's demonstration, however, was much longer and more complicated than the concise and elegant demonstration given by Torricelli. Torricelli assumed *a priori* that the circumferences of circles are to one another as their diameters. He also assumed that the circle with center A and radius AB is rectified, i.e. its circumference is equal to the straight line BC, tangent to the circle. He then considered (cf. Figure 3.4) a concentric circle of diameter IO and a parallel line (IL) to (BC). Let us write $\overset{\frown}{BD}$ to designate the circumference of the circle. We have $\overset{\frown}{BD}:\overset{\frown}{IO} = BA:AI$ and $BA:AI = BC:IL$. We then get $\overset{\frown}{BD}:\overset{\frown}{IO} = BC:IL$ and, "*permutando*" [!], $\overset{\frown}{BD}:BC = \overset{\frown}{IO}:IL$. Since $\overset{\frown}{BD} = BC$ by hypothesis, finally $\overset{\frown}{IO} = IL$. The same is true for all concentric circles filling the given one BD, thus 'the whole [circular] lines' which constitute the circle BD are equal to the "whole straight lines" of triangle ABC, i.e. all the lines included in the triangle ABC. By means of the method of indivisibles, Torricelli thus easily derived the desired conclusion.

Among the promoters of the new method of indivisibles was John Wallis (1616–1703), who held the Savilian Chair of Geometry at Oxford. In his *Arithmetica Infinitorum* (Arithmetic of Infinitesimals, 1656), Wallis adopted an arithmetical and numerical point of view, in contrast with the classic geometrical soil of the method, even though he became aware of indivisibles through Torricelli's work. He determined sums of power series and deduced an expression for π by means of successive interpolations. Wallis's approach also had its roots in the British tradition of numerical approximation, represented, for instance, by the studies on logarithms of John Napier (1550–1617) and Henry Briggs (1556–1630). Wallis used the method of "incomplete induction," which allowed him to assume that a property established for a finite number of terms in a series also held for an infinite number of terms. Newton would later

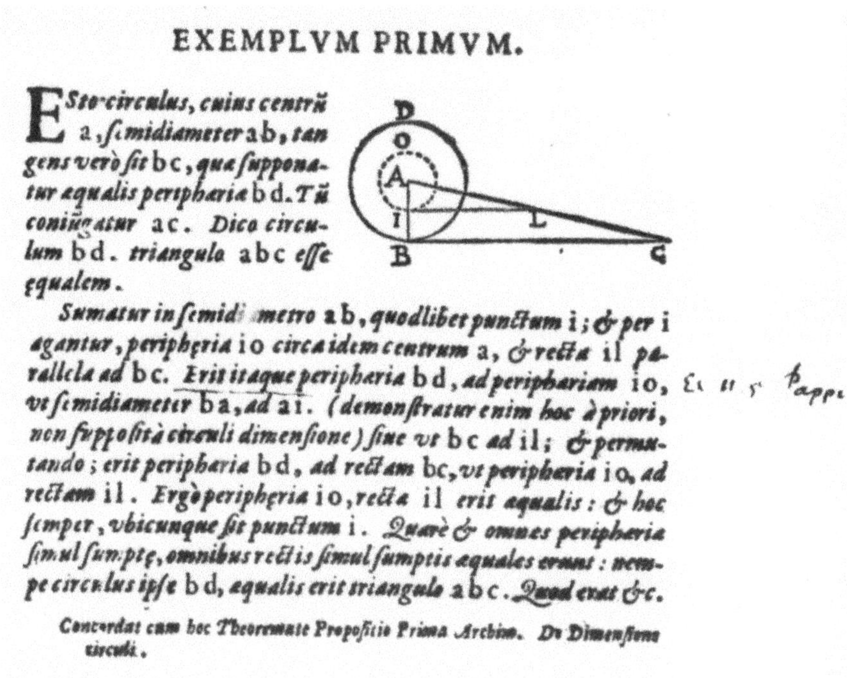

EXEMPLVM PRIMVM.

Esto circulus, cuius centrũ a, semidiameter ab, tangens verò sit bc, quæ supponatur æqualis peripheriæ bd. Tũ coniũgatur ac. Dico circulum bd. triangulo abc esse æqualem.

Sumatur in semidi metro ab, quodlibet punctum i; & per i agantur, peripheria io circa idem centrum a, & recta il parallela ad bc. Erit itaque peripharia bd, ad periphariam io, vt semidiameter ba, ad ai. (demonstratur enim hoc à priori, non supposità circuli dimensione) sive vt bc ad il; & permutando; erit peripharia bd, ad rectam bc, vt peripheria io, ad rectam il. Ergò peripheria io, recta il erit æqualis: & hoc semper, vbicunque sit punctum i. Quarè & omnes peripheriæ simul sumptę, omnibus rectis simul sumptis æquales erunt: nempe circulus ipse bd, æqualis erit triangulo abc. Quod erat &c.

Concordat cum hoc Theoremate Propositio Prima Archim. De Dimensione circuli.

FIGURE 3.4. Torricelli, “De Dimensione Parabolae,” *Opera Geometrica* (1644: 95). Max Planck Institute for the History of Science, Berlin.

draw some of his important early discoveries, for instance his invention of the general binomial theorem, from his reading of Wallis.

One of the most important and certainly the greatest mathematical innovation of the early modern period, namely the independent invention of the calculus by Leibniz and Newton, resulted from many developments throughout the seventeenth century. These included algebraic and kinematic methods for finding tangents as well as new methods of quadrature involving indivisibles and infinitesimals. Even if the works of Newton and Leibniz were the first to exhibit *algorithms* for systematizing and combining various diverse infinitesimal methods, a surfeit of earlier works provided “foreshadowings of the calculus” from the 1630s up to the 1660s (Baron [1969] 2004; Møller Pedersen 1980 [2000]; Stedall 2008). For finding tangents, one could cite Fermat, Roberval, and Barrow; for quadratures, Pascal and Wallis. Methods and mathematical results sometimes circulated among mathematicians through letters or meetings. In other cases, mathematicians working independently produced *bona fide* identical or nearly equivalent mathematical results. Controversies, naturally enough, often followed, as was the case at the very beginning with the methods used by Descartes and Fermat for finding tangents. And this would also be the case at the end of the century with the debates with regard to the respective "rights of invention" of Newton and Leibniz.

Roberval's kinematic method based on the "composition of movements," which he applied in constructing tangents to geometrical *and* transcendental curves, offers an impressive example of the circulation of methods described above. Roberval's idea was to consider a curve as the path of a moving object. The tangent at a given point was therefore the direction of the motion generating the curve at a particular instant of time. This kinematic method was developed in a manuscript from the 1640s, written up based on Roberval's lectures by his pupil François Bonneau Du Verdus (1621–75). This text only appeared in print form in 1693, posthumously published in Roberval's works, but various manuscript copies had long before circulated within learned European circles. Newton, who probably had access to Roberval's ideas, incorporated analogous arguments and elaborated on this very kinematic method in his manuscripts from 1665. He used it to find tangents to transcendental curves, which had been rejected by Descartes from geometry, including spirals, cycloids, the quadratrix, and logarithmic curves.

The method of tangents that Barrow developed in his *Lectiones Geometricae* (Geometrical Lectures, 1669) can be seen as occupying an intermediate place between Fermat's method (invented *c.* 1629) and that of Leibniz. On any relevant portion of a curve, Barrow introduced two infinitely near points. He then considered the triangle formed by infinitesimal increments e and a of the abscissa x and the ordinate y, respectively. Such a figure, which Leibniz would later call the "characteristic triangle," was "similar" to the triangle formed by the subtangent t, i.e. the segment on the axis intercepted by the tangent and the ordinate y, the ordinate, and the tangent; it thus follows that $\frac{y}{t}=\frac{a}{e}$. By using this proportion together with the curve equation, and by neglecting infinitesimally small quantities, Barrow was able to determine the subtangent to the curve and thus to provide a geometrical construction for the tangent. Here, the ratio $\frac{a}{e}$ was used as a tool rather than as an object of study per se.

Let us consider this method by constructing the tangent to a point on a parabola, a classical result already established by Apollonius in his *Conics* (Prop I.33; cf. Figure 3.1). If $B\ (x, y)$ and $b\ (x + e, y + a)$ are two infinitely near points on a parabola with equation $y^2 = px$, and if $PT = t$ is the subtangent (cf. Diagram 3.1), then since the triangle Bbc is "similar" to the triangle TPB, we get:

$$\frac{PB}{PT}=\frac{cb}{cB} \text{ or } \frac{y}{t}=\frac{a}{e} \qquad (1).$$

On the other hand, since b lies on the parabola,

$$(y + a)^2 = p(x + e), \text{ viz. } y^2 + 2ay + a^2 = px + pe.$$

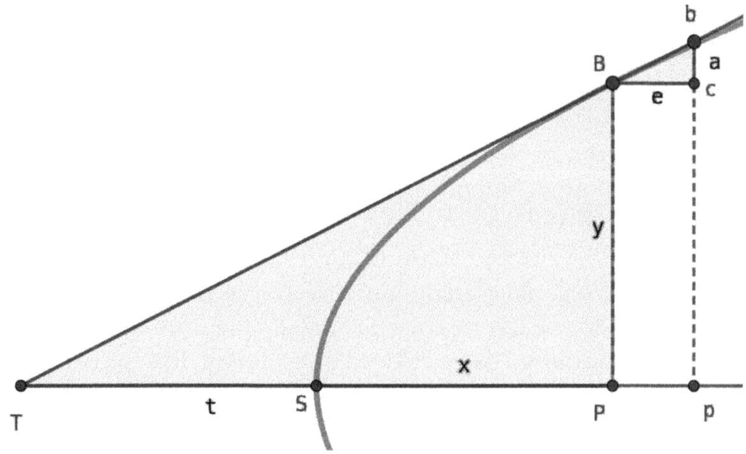

DIAGRAM 3.1. Barrow's method applied to the tangent of parabola. Courtesy of Sébastien Maronne.

Now we can remove the equal quantities y^2 and px (since B belongs to the parabola) and neglect the infinitely small a^2 to thus obtain:

$$2ay = pe \text{ or } \frac{a}{e} = \frac{p}{2y} \qquad (2).$$

Finally, we conclude from (1) and (2) that:

$$\frac{y}{t} = \frac{p}{2y} \text{ and the subtangent } t = 2\frac{y^2}{p} = 2x.$$

This method was already very close to the one invented a bit later by Leibniz, and was perceived as such by his contemporaries. Where Barrow used small increments, e and a, of the abscissa x and the ordinate y, Leibniz introduced differentials cleverly labeled dx and dy. This highly efficient notation was a great step forward as it favored the operational aspects of the calculus.

THE INVENTION OF THE CALCULUS BY NEWTON AND LEIBNIZ

It is well known today that Leibniz and Newton independently invented the calculus, even though this had long been a disputed issue. Newton conceived his fluxional calculus in 1664–6, but went on developing his ideas during the following years, although his researches only began to be publicized during the early eighteenth century. For his part, Leibniz devised a somewhat different form of the differential and integral calculus a decade later, in 1675, though he published his version in the mid-1680s in two articles in the *Acta Eruditorum* long before Newton's work would appear in print (Baron [1969] 2004; Bos 1980 [2000]; Stedall 2008). A famous and hard-fought priority controversy concerning the invention of the calculus thereafter ensued, during which from 1708 to 1730 supporters on both sides threw verbal abuse at the other (Hall 1980; Bertoloni Meli 1993).

In the works of Leibniz and Newton, curves and variable quantities, which were the main objects under study, were conceived in quite different ways. According to the latter, curves were generated by the continuous motion of a point. The quantities curves connected with the motion, like the coordinates x, y of the point, varied with respect to time. Newton called these "fluents," whereas "fluxions" were the instantaneous rates of change of these fluents, in his masterpiece, the so-called *De methodis serierum et fluxionum* (Method of Fluxions and Infinite Series), of which he had completed a first draft in 1671 but which was only published posthumously in an English translation in 1736. It was later, most probably in the early 1690s, that Newton invented the famous "dot-notation" $\dot{x}$, $\dot{y}$ to designate fluxions of the fluents x, y. More importantly, Newton *introduced an algorithm* to determine the ratio $\frac{\dot{y}}{\dot{x}}$ between the fluxions, which gives the slope of the tangent. He derived this from the relationship between the fluents x, y as given by the equation of the curve.

Newton's invention of the calculus took some of its inspiration from earlier contributions. At the beginning of his research, he benefited from Hendrik van Heuraet's (1634–*c.* 1660) study on the rectification of curves in *Epistola de transmutatione curvarum linearum in rectas* (Letter on the Transformation of Curves into Straight Lines, 1659). Newton also read Johann Hudde's (1628–1704) works on maxima and minima and the theory of equations. The works of these Dutch mathematicians had been published in the second Latin edition of Descartes's *Geometry* (1659–61), edited by van Schooten. Newton also studied Wallis's *Arithmetic of Infinitesimals* (1656) and enjoyed mathematical exchanges with Barrow, whom he later succeeded as Lucasian Professor of Mathematics at Cambridge in 1669.

Although publication of Newton's research on the calculus was long delayed, his research circulated among a small circle of mostly British mathematicians, which included Barrow, Wallis, James Gregory (1638–75), and John Collins (1625–83). These scholars were thus familiar with Newton's 1669 manuscript entitled *De Analysi per aequationes numero terminorum infinitas* (On the Analysis by Means of Equations with an Infinite Number of Terms). This dealt with rectifications and quadratures by means of series expansions and was originally written in response to the publication one year earlier of Nikolaus Mercator's (1620–87) *Logarithmotechnia*, in which the German mathematician gave an infinite series for the logarithm. The tract *De Analysi* would be printed much later in 1711. It seems that Newton, having become more and more fascinated by ancient geometry, preferred to leave his great discoveries undisclosed rather than to expose himself to criticism because his research did not meet the ancient standards of proof and rigor. Today, one can gain a precise idea of Newton's complex invention process by examining his mathematical manuscripts. These were edited by Derek-Thomas Whiteside and published from 1967 to 1981 by Cambridge University Press in the eight volumes of the *Mathematical Papers of Isaac Newton*.

In sharp contrast to Newton, Leibniz conceived of curves as polygons with infinitely many sides. The "differentials" dx, dy were the infinitely small differences between two successive values of the abscissa and the ordinate x, y. From the characteristic triangle (Diagram 3.1), one easily derived the proportion $\frac{dy}{dx} = \frac{y}{t}$, where t is the subtangent, the very same proportion that appears in Barrow's method. On the other hand, by considering equidistant ordinates y_k, where $y_{k+1} - y_k$ was taken equal to an *infinitely small unit*, the quadrature of the curve was obtained as the *sum* of the ordinates $\Sigma_k y_k$ and the slope of the tangent as the *difference* $y_{k+1} - y_k$ between two consecutive ordinates. Such reasoning allowed Leibniz to recognize that the calculation of quadratures and tangents involved inverse procedures that were analogous to addition and subtraction.

Leibniz invented his differential and integral calculus in October-November 1675 when he was in Paris. He then introduced the symbol $\int$ to replace the notation for "all the lines" used with Cavalieri's indivisibles in quadrature problems as well as the differential symbol dx. Leibniz was indeed very much interested in the development of a general symbolic language in which sound reasoning would be grounded in rules for combining suitable symbols. Leibniz's invention of calculus was also rooted in a fusion of ideas and methods found in the works of Cavalieri, Descartes, Pascal, and Barrow. He received his training in higher mathematics in Paris during the fruitful years from 1672 to 1676 when he studied under the guidance of the Dutch mathematician Christiaan Huygens (1629–95).

In October 1684, Leibniz published his differential algorithm in a famous article of seven pages, *Nova methodus pro maximis et minimis, itemque tangentibus* (A New Method for Maxima and Minima, and for Tangents) (cf. Figure 3.5). It should be noted that Leibniz decided for the sake of rigor *not* to define differentials as infinitely small but as small *finite* quantities. The differential dx designated a line segment taken arbitrarily and the differential dy was defined as the line segment satisfying the proportion $\frac{dy}{dx} = \frac{y}{t}$. In this article, Leibniz stated the general rules for differentiation, the conditions $dv = 0$ for an extremum, and the condition $d\,dv = 0$ for an inflection point. Two years later, in June 1686, he published his integral calculus in an article of nine pages

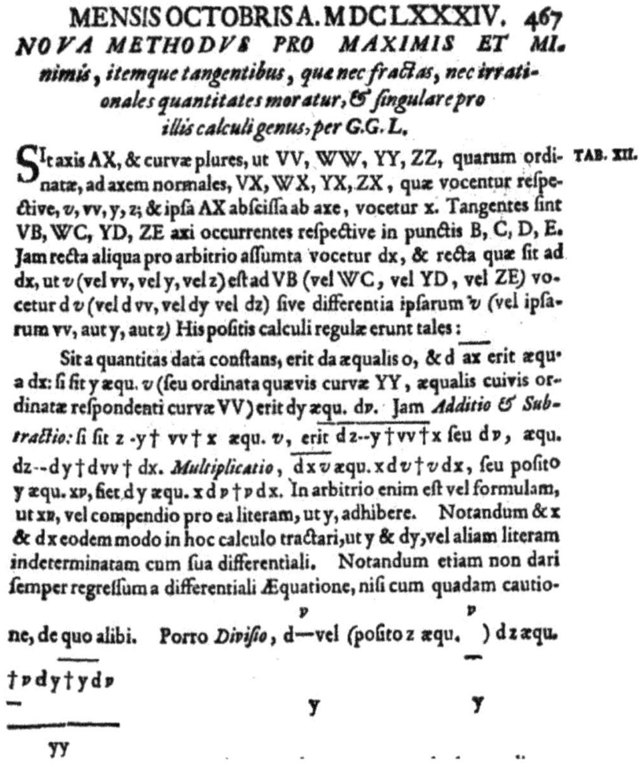

MENSIS OCTOBRIS A. MDCLXXXIV. 467

NOVA METHODUS PRO MAXIMIS ET MInimis, itemque tangentibus, quae nec fractas, nec irrationales quantitates moratur, & singulare pro illis calculi genus, per G.G. L.

Sit axis AX, & curvæ plures, ut VV, WW, YY, ZZ, quarum ordinatæ, ad axem normales, VX, WX, YX, ZX, quæ vocentur respective, v, vv, y, z; & ipsa AX abscissa ab axe, vocetur x. Tangentes sint VB, WC, YD, ZE axi occurrentes respective in punctis B, C, D, E. Jam recta aliqua pro arbitrio assumta vocetur dx, & recta quæ sit ad dx, ut v (vel vv, vel y, vel z) est ad VB (vel WC, vel YD, vel ZE) vocetur dv (vel dvv, vel dy vel dz) sive differentia ipsarum v (vel ipsarum vv, aut y, aut z) His positis calculi regulæ erunt tales: TAB. XII.

Sit a quantitas data constans, erit da æqualis o, & d ax erit æqu. a dx: si sit y æqu. v (seu ordinata quævis curvæ YY, æqualis cuivis ordinatæ respondenti curvæ VV) erit dy æqu. dv. Jam *Additio & Subtractio:* si sit z -y † vv † x æqu. v, erit dz--y † vv † x seu dv, æqu. dz--dy † dvv † dx. *Multiplicatio*, dxv æqu. xdv † vdx, seu posito y æqu. xv, fiet dy æqu. xdv † vdx. In arbitrio enim est vel formulam, ut xv, vel compendio pro ea literam, ut y, adhibere. Notandum & x & dx eodem modo in hoc calculo tractari, ut y & dy, vel aliam literam indeterminatam cum sua differentiali. Notandum etiam non dari semper regressum a differentiali Æquatione, nisi cum quadam cautione, de quo alibi. Porro *Divisio*, d $\frac{v}{y}$ vel (posito z æqu. $\frac{v}{y}$) dz æqu. $\frac{\dagger v dy \dagger y dv}{yy}$

FIGURE 3.5. The differentiation rules in Leibniz, "Nova Methodus pro Maximis et Minimis, itemque tangentibus," *Acta eruditorum* (1684: 467). Niedersächsische Staats- und Universitätsbibliothek Göttingen.

entitled *De geometria recondita et analysi indivisibilium atque infinitorum* (On a Hidden Geometry and Analysis of Indivisibles and Infinitesimals). There he presented and explained the integral sign ∫ and its use in quadrature problems.

Despite the rather abstruse presentation of Leibnizian calculus in these two brief articles in *Acta Eruditorum*, it spread rather quickly within the mathematical community. Already in the 1690s, the brothers Jacob Bernoulli (1655–1705) and Johann Bernoulli (1667–1748) applied it to solve problems in mathematical physics (cf. Chapter 3 in Volume 4). Moreover, the handbook published in 1696 by Guillaume de l'Hôpital (1661–1704), *Analyse des Infiniment petits pour l'intelligence des lignes courbes* (Analysis of Infinitely Small Quantities for the Understanding of Curved Lines), was written based on lessons taught by Johann Bernoulli. In the decades that followed, it greatly contributed to the dissemination of Leibnizian calculus. As evidenced by the title of L'Hôpital's tract, curves had by this time become a prime object of study in mathematics.

Finally, what were the main mathematical ingredients that entered into the new calculus? One was the *general binomial theorem* discovered by Newton during the winter of 1664-1665, which gives the (infinite) power series expansion of $(a + x)^r$ for r rational, thereby generalizing the well-known (finite) binomial expansion of $(a + x)^n$ for a positive integer n. Another was the *characteristic triangle* (Diagram 3.1) found in 1675 by Leibniz and which proved to be a useful tool in both quadrature and tangent problems. A third key ingredient came with the discovery now known as the *fundamental theorem of calculus*, which points to how tangency and quadrature problems are related by means of the inverse operations of differentiation and integration. But, above all, it was the development of systematic *algorithms* and adequate notations that contributed to the birth of calculus.

CONCLUSION

As we have seen, Newton attached much importance to an ideal of rigor inherited from the ancients. While he was rightly convinced by the substantial innovations he made in mathematics, he nevertheless saw these as falling within a long-lasting tradition shaped by his predecessors. In the same vein, he cautiously used *geometric* limit procedures in the *Principia*, which he wrote in a classical synthetic style. Whether or not he used infinitesimal calculus in his masterwork would remain a disputed question between Newtonians and Leibnizians, however. In this sense, Newton's *Principia* marks a fitting end for our journey through early modern mathematics betwixt tradition and innovation.

The heirs of Newton and Leibniz would be much less reluctant to use the powerful tools of infinitesimal calculus. They questioned to a far lesser extent such use, nor did they tend to ponder the very nature of infinitesimal and infinitely large quantities. We thus enter here into a new mathematical period, during which the calculus proved to be a powerful tool for mathematicians who combined finite as well as infinite methods to solve problems in both mathematics and physics. These developments gradually led to the accession of a new queen of mathematics: the notion of function (cf. Chapter 3 in Volume 4). This new concept would eventually dethrone the geometric curve and go on to become the principal object of study in the next century.

CHAPTER FOUR

Mathematics and Worldviews

Mathematics as a Touchstone for "Certainty"

DAVID RABOUIN

In his semi-clandestine *Historiettes*, the French writer Tallemant des Réaux (1619–92) describes the death of the Prince of Orange and *stadtholder* of the Dutch Republic, Maurice of Nassau (1567–1625), in this colorful way:

> One thing is quite notable about the end of this great man. Being at the end, he summoned a minister and a priest, and made them dispute about religion; and after hearing them long enough he said: 'I see that there is nothing certain but mathematics'. And having said this, he turned to the other side, and exhaled. (Tallemant des Réaux, *Historiettes*, Paris: A. Levavasseur, 1834–5: vol. 1, 328)

That this narrative is not merely anecdotal is confirmed by a similar dialogue, occurring in Molière's famous play *Dom Juan* of 1665 (Act 3, Scene 1):

> **Sganarelle** That is a man I will have a lot of trouble converting. But tell me, as one must believe something: What do you believe?
>
> **Dom Juan** What do I believe?
>
> **Sganarelle** Yes.

Dom Juan I believe that two and two makes four, Sganarelle, and that four and four makes eight.

Sganarelle A fine credo indeed! Your religion is therefore, as I see it, Arithmetic?

These two quotes directly lead into the topic of this chapter as they indicate a profound change in the way mathematics was integrated in individual worldviews. From a specialized, almost esoteric field of knowledge, cultivated by scholars or technicians, it spread in Western elite culture, ascending to a position where belief in mathematics could challenge religion. Even facing death, Maurice of Nassau had no doubt, just as Dom Juan was sure: they would believe in nothing else but mathematics. But the position they stood for is even stronger: if "there is nothing certain but mathematics," this also means that mathematics becomes the standard for *any kind* of knowledge. This chapter will assess this new cultural position of mathematics. It takes the standard of "certainty" as a travel guide and provides a historical account of its emergence as a touchstone for knowledge. Since this context is epistemological, particular emphasis will be placed on the ramifications of this for the relationship between mathematics and philosophy (which, at the time, still encompassed "natural philosophy").

First, we will briefly recapitulate a familiar narrative closely linked to the so-called "Scientific Revolution." It relates this rising standard of "certainty" to a kind of revenge of Plato against Aristotle (whose views dominated natural philosophy in the Middle Ages, especially in European universities). However, this familiar narrative will be challenged in this chapter—I argue that the question of certainty did not suddenly emerge along with the rediscovery of Plato and his plaidoyer for mathematics in the Renaissance. It was a precise technical problem (known as the *quaestio de certitudine mathematicarum*) originating in the late scholastic philosophy (i.e., the Aristotelian doctrine taught in European universities). Moreover, I will show that this debate had a deep influence on the conceptions of mathematics formulated in the seventeenth century. Seen from this perspective, the development of mathematics into an epistemological gold standard travels a rather continuous path from the end of the fifteenth century to the seventeenth century (Mancosu 1996: ch. 1). With this in mind, we will turn to the specific topic of infinity, because it offers an excellent example of how the debate about the certainty of mathematics is a game changer. Finally, we will turn to the new epistemological problems which accompanied the introduction of new mathematical entities such as infinitesimals or imaginary quantities. To conclude, we will also look at the way in which the new standard of the certainty of mathematics encountered other cultural models and related to them.

PLATO'S REVENGE?

The relationship between mathematics and worldviews in the early modern period appears as closely linked to the radical cultural change known as the "Scientific Revolution" and the shift occurring with the publication of Nicolaus Copernicus's (1473–1543) *De revolutionibus orbium coelestium* (On the Revolutions of the Heavenly Spheres) in 1543. By expelling the earth from the center of the world and by seeing the universe as a geometrical space, infinite and silent, a new era began (Koyré 1957). Mathematics suddenly became constitutive, although not without resistance, of the way the so-called Western culture looked at "Nature"—to the point of becoming a touchstone by which other cultural frameworks are still distinguished by anthropologists from this emerging "naturalism" (Descola [2005] 2013). Although mathematics played a central role in many other cultural contexts and could serve to describe natural phenomena (typically optical and astronomical facts), it had never before played such a widespread role, impregnating all realms of thought, including not only natural philosophy, but also theology, law, politics, or ethics, and even reaching out, as we have seen, to the stages of the theatres and the aristocratic salons.

Mathematics began to be the language for any truthful description of Nature and, it is thought, the very standard of what truth meant. As expressed by Galileo Galilei (1564–1642) in an oft-quoted passage from his *Saggiatore* (Assayer) of 1623:

> Philosophy is written in this grand book, the universe, which stands continually open to our gaze. But the book cannot be understood unless one first learns to comprehend the language and read the letters in which it is composed. It is written in the language of mathematics, and its characters are triangles, circles, and other geometric figures without which it is humanly impossible to understand a single word of it; without these, one wanders about in a dark labyrinth. (Drake 1957: 237)

This claim is reminiscent of the anecdote about Maurice of Nassau's death. While the "book of nature" was usually interpreted, before that time, in subjugation to the other "Book," the Holy Scriptures, it is now boldly claimed that mathematics is its self-sufficient language. The change is even more striking when one keeps in mind that, in the previous period, European higher education was still dominated by Aristotelian traditions. Indeed, for Aristotle and his commentators, natural philosophy and mathematics had to remain separate domains of knowledge—in a clear opposition to his master Plato who, in the *Timaeus*, presented a mathematized structure of the world and praised mathematics as an entranceway to the highest form of knowledge, the "world of ideas."

In the sixteenth and seventeenth centuries, after a long time during which Aristotelian philosophies, mixed with various forms of religious beliefs, had been standard in European universities and barred the road to a mathematization of nature, it was high time, as Alexandre Koyré (1892–1964) put it, for "Plato's revenge" (Koyré [1939] 1966: 290).

This gradual evolution seems to have been supported by the rapid development of a mathematized "natural philosophy" as practiced by authors, besides Galileo, such as Johannes Kepler (1571–1630), Christiaan Huygens (1629–95) or Isaac Newton (1642–1727) (cf. Chapter 5). It seems also to have been made possible by the invention of new mathematical tools such as symbolic algebra, analytical geometry, and infinitesimal calculus. A new universe of mathematics and its powers emerged at the time, which could apparently be understood simply by reading its formulae (a belief which certainly does not hold true, even as a retrospective illusion, for mathematics coming from other cultures or previous periods). A new mathematics which also now appears, or so we think, as a purely symbolical language.

This process of mathematization shaped our "modernity" profoundly and has been seen as permeating every part of so-called "Western culture." Western capitalism and imperialism were definitively marked by the increasing power of quantification. As later critiques would put it, still in relation to the supposed revenge of Plato:

> The mythologizing equation of Forms with numbers in Plato's last writings expresses the longing of all demythologizing: number became enlightenment's canon. The same equations govern bourgeois justice and commodity exchange. [. . .] Bourgeois society is ruled by equivalence. It makes dissimilar things comparable by reducing them to abstract quantities. (Horkheimer and Adorno [1944] 2002: 4)

A similar picture has been drawn by Michel Foucault (1926–84) with respect to the development of mercantilism in chapter VI of *The Order of Things* (Foucault [1966] 1970). This does not seem to be too much of an exaggeration, if we consider that the development of symbolic algebra was closely connected in the sixteenth century to the development of modern accounting methods and, later, with the birth of modern state planification (think of the role of Colbert in France), or that economics emerged as a new field in the early modern period, or that some of the first theories of "Natural law" were conceived "in the manner of the geometers" and based on a proto-calculus of utilities (i.e. the idea that human interactions are based on unconscious "computations" of what is best for each individual).

From a philosophical point of view, our period is marked by: (1) the progressive rejection in Europe of the obscurities of Aristotelian philosophies ("scholasticism") by Renaissance humanists, and (2) the development of the great "rationalist" systems attached to the names of René Descartes (1596–1650), Baruch Spinoza (1632–77), Nicolas Malebranche (1638–1715), and Gottfried Wilhelm Leibniz (1646–1716). One important feature of the first trend was the rediscovery of ancient knowledge, in particular Greek mathematics, and putting it in print, for instance, Euclid's *Elements* published in Venice in 1482 (cf. Chapter 7). One important feature of the second was to posit mathematics as the standard for evaluating the certainty of any discourse. While it would be wrong to think that these processes unfolded without resistance and criticisms, the very nature of philosophical debates assuredly changed, as did the standard for truth. New problems typical of our philosophical "modernity"—such as the relationship between the thinking subject and its body, the exact nature of its freedom, the way in which human beings could or could not render themselves as "masters and possessors of nature"—emerged at the time in connection with claims about mathematical certainty. Moreover, these problems and debates are still at the center of many of our contemporary interrogations and seem to accompany naturally the development of "Modern Science."

Yet, a better knowledge of the development of early modern science and philosophy in their relationship to mathematics has shown that this harmonious picture is largely a mythical reconstruction and lacks important nuance. Let us start by recalling that the so-called "Copernican Revolution," far from being a sudden radical change of worldview, drew out over a long period of time, in part because very few people were able to grasp its content. One should always keep in mind that the formal condemnation of heliocentrism by the Catholic Church, as emblematic as it may seem, did not occur before 1616. This was more than seventy years after the publication of Copernicus's book. In the same manner, the idea that symbolic algebra and infinitesimal calculus paved the way for a completely new understanding of the relationship between mathematics and nature, although true, neglects the very long time it took for these techniques to become widely used. One just has to open the works of Galileo or Kepler (cf. Chapter 5), but also, more surprisingly of Descartes, Newton or Leibniz, to see that they still gave a rather prominent role to the geometric "theory of proportions" and newer methods played a smaller role. Moreover, as we will see, it is simply not true that mathematics had not been considered as a standard of truth by supporters of Aristotle and then was suddenly reappraised under a "Platonist" influence in the Renaissance. By reading our authors, it sometimes even seems to be quite the contrary.

ON MATHEMATICAL CERTAINTY

To sketch a more nuanced picture than the one drawn in the preceding section, let us start with an intriguing puzzle. In a famous booklet published in 1547 under the title *Commentarium de Certitudine Mathematicarum Disciplinarum* (Commentary on the Certainty of the Mathematical Sciences), the Italian humanist Alessandro Piccolomini (1508–79) wrote that he had learned at university, where the teaching of Aristotle still dominated, that mathematics was at the *highest place* when it came to certainty. Indeed, Aristotle was taught in the universities through many mediations, a very important one being that of Arabic commentators such as Averroes (the Latin name for Ibn Rushd, 1126–98). Averroes assigned a much higher status to mathematics than Aristotle had. Moreover, he held a prominent place in the teaching of philosophy and natural philosophy in the universities of Northern Italy (Rose 1975).

Piccolomini explains that he only came to realize that one could criticize and doubt the certainty of mathematics when reading the commentary of Proclus on Euclid's *Elements* (first published in Greek 1533 and translated in Latin in 1560, see Figure 4.1). Proclus, who lived in the fifth century CE, was the head of the *Akademia*, the late descendant of Plato's school in Athens, created eight centuries before. A typical "Platonist" is thus mentioned as an authority in support of the *uncertainty* of mathematics and against the "Aristotelians" who were supporting its highest certainty. The exact opposite of the picture Koyré, and so many authors after him, thought as *constitutive* of our modernity.

This situation is of particular interest, because one can show that authors such as Galileo belong to this very particular milieu of North Italian humanists (De Pace 1993). This already changes the way we assess "Plato's Revenge" quite drastically. Whereas it is tempting to see the constitution of our "modernity" as a fight between supporters of Plato, defending the primacy of mathematics, and supporters of Aristotle, refusing such a model, a closer look at the sources indicates that this kind of staging is a mythical reconstruction. It says more about our own belief in the grandiose Eurocentric narrative of the constitution of our "modernity" (against preceding "dark ages") than about the way the actors themselves experienced their relation to mathematical knowledge.

Without going into any details of the debate tackled by Piccolomini, it is important to realize that the question about the certainty of the mathematical sciences he raised, the so-called *Quaestio de certitudine mathematicarum*, was not limited to a simple rejection or defense of the certainty of the mathematical sciences. The aim of the question was rather what *kind of certainty* they could provide. Its core was a distinction established by Aristotle between different types of proofs: a proof establishing a mere fact (in Greek: *to oti*; in Latin: *quia*) and a proof giving the reason of this fact (*to di oti/propter quid*). A common

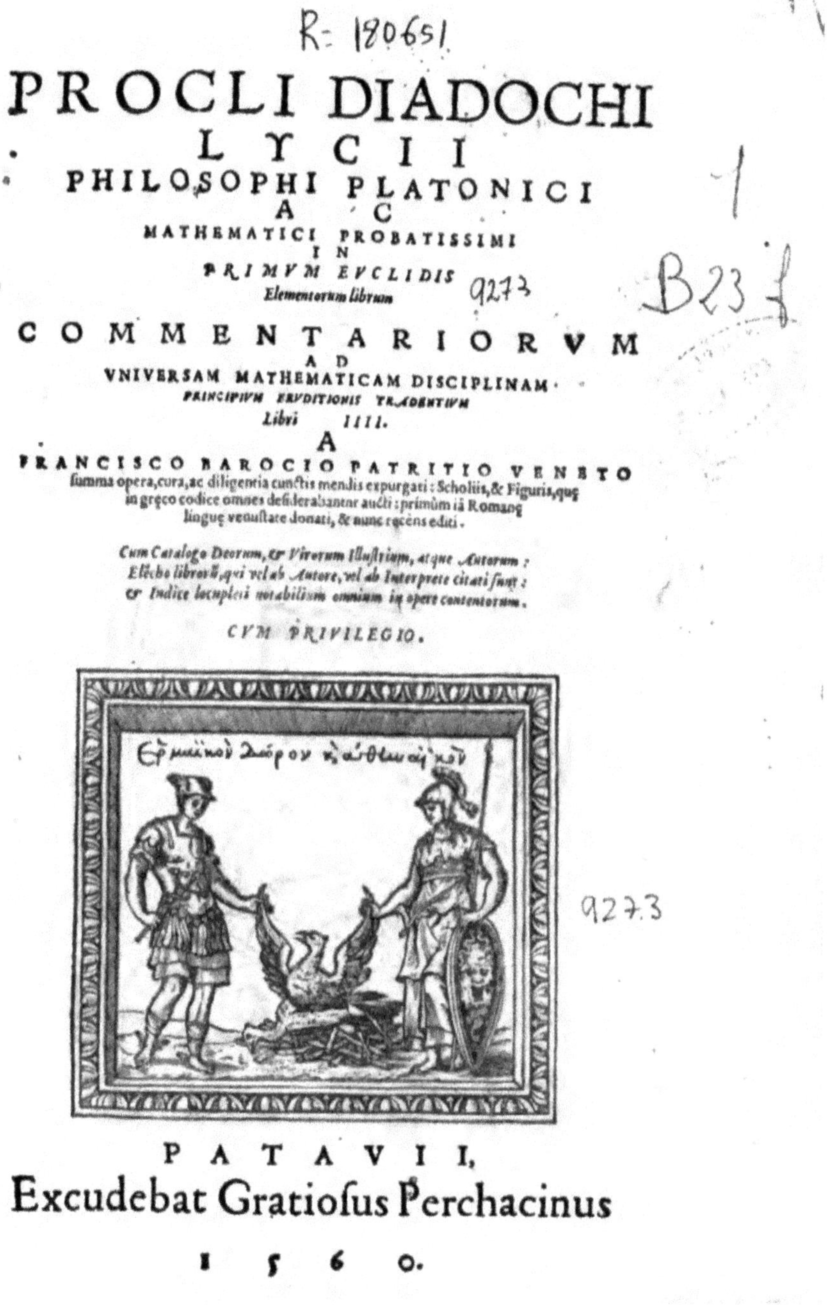

PROCLI DIADOCHI
LYCII
PHILOSOPHI PLATONICI
AC
MATHEMATICI PROBATISSIMI
IN
PRIMVM EVCLIDIS
Elementorum librum
COMMENTARIORVM
AD
VNIVERSAM MATHEMATICAM DISCIPLINAM
PRINCIPIVM ERVDITIONIS TRADENTIVM
Libri IIII.
A
FRANCISCO BAROCIO PATRITIO VENETO
summa opera,cura,ac diligentia cunctis mendis expurgati: Scholiis,& Figuris,quę in grȩco codice omnes desiderabantur aucti: primùm iã Romanę linguę venustate donati, & nunc recèns editi.

Cum Catalogo Deorum, & Virorum Illustrium, atque Autorum: Elẽcho librorũ,qui vel ab Autore,vel ab Interprete citati sunt: & Indice locupleti notabilium omnium in opere contentorum.

CVM PRIVILEGIO.

PATAVII,
Excudebat Gratiosus Perchacinus
1560.

FIGURE 4.1. The Latin translation of Proclus's commentary on Euclid by Francesco Barozzi: *Procli Diadochi Lycii In primum Euclidis* (1560). Courtesy of Internet Archive.

example of the first was given by the inference from the fact that planets are non-twinkling bodies to the fact that they should henceforth be near the earth. An example of the second was given by the inference from the fact that planets are near the earth to the fact that they are non-twinkling bodies. The first proceeds from the effect to the cause, the second from the cause to the effect. According to Averroes, the highest degree of certainty would be given by a combination of these two features. He called this type of proof "most powerful" (translated by the Latin authors as *potissima*). A prototype of this kind of proof was supposedly provided by mathematics in which one concludes by giving both the fact *and* the reason of the fact at the same time. This is why Averroes posited mathematics at the highest rank when it comes to "certainty" and this opinion, as recalled by Piccolomini, was largely shared by Latin masters of the Schools.

After studying Proclus, however, Piccolomini realized that this belief was far from obvious. This should not surprise us too much since Proclus's commentary on Euclid had not been known in the medieval period (Latin or Arabic) and it assuredly contains a vast repertoire of criticisms raised against the *Elements*. Amongst these criticisms was a famous attack against *Elements* I.32, the emblematic proposition stating that the sum of the interior angles of a triangle is equal to two right angles.

Indeed, for the conduct of the proof, Euclid extends one of the sides, draws a parallel to the other (in dotted lines in Diagram 4.1) and concludes by relying on equalities between interior and exterior angles. Proclus recalled that this proof suffered from the fault of relying on auxiliary constructions. Taken in and of itself, a triangle does not have lines "external" to it. Yet, a proper demonstration, proceeding "from the cause," should derive a property from the nature or "essence" of the object. For this reason, Euclid's proof could not be considered perfect. Moreover, after recalling this criticism, Proclus provided an alternative proof, without recourse to external constructions and which he praised because it enabled one to see the "genesis" or reason for the fact demonstrated (hence the later designation of this kind of proof as "genetic").

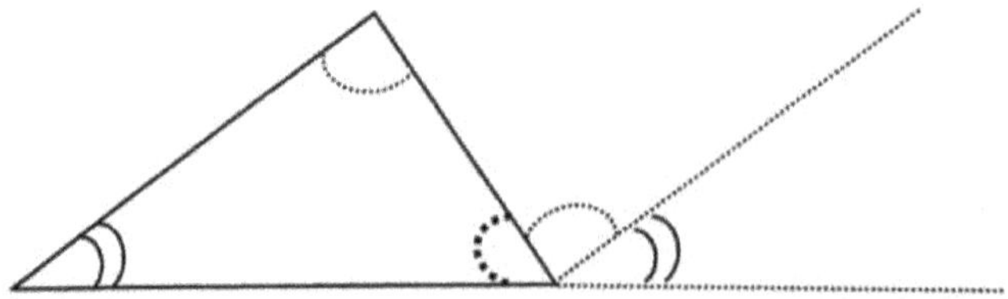

DIAGRAM 4.1. A diagram for Euclid's *Elements* I.32. Courtesy of David Rabouin.

Critics of Piccolomini immediately claimed that he had not understood Proclus properly. Indeed, Proclus's aim was not to doubt Euclid's reasoning, but to identify a kind of proof. Among the first of these scholars was Francesco Barozzi (1537–1604), who provided a translation of Proclus's treatise into Latin in 1560 and gave lectures on it in the University of Padua. Others, such as the Portuguese Jesuit Benito Pereira (1535–1610), professor at the *Collegio Romano* (the central Jesuit College established in Rome), sided with Piccolomini. The unfolding vivid debate extended throughout the seventeenth century and was taken up in many new institutions where the teaching of mathematics was beginning to be systematized all over Europe. From the College of Santo Antão in Lisbon to the Academy of Vilnius, including the Collège Royal in Paris, the University of Ingolstadt and various Protestant universities such as Strasbourg or Jena, the question of the certainty of mathematics agitated many brilliant minds of the time.

THE DEBATE ON CERTAINTY IN THE SEVENTEENTH CENTURY

Discussions of the "certainty" of the mathematical sciences can be found throughout the seventeenth century, for example in the works of Tommaso Campanella (1568–1639), Pierre Gassendi (1592–1655), Robert Boyle (1627–91), amongst other authors (Sergio 2006). "Experimental philosophers" like Boyle, who constituted the vast majority of scholars associated with the Royal Society, founded in 1660 in London, were very critical of a use of mathematics which would not be regulated by experimental practice. Their reasons were often reminiscent of what had already been advocated by late scholastic thinkers: mathematics, although useful, does not exhibit the "real" causal processes on which any knowledge of nature should be built. Even in the *Collegio Romano*, the great campaign launched by Christopher Clavius (1538–1612) to put mathematics at the center of the teaching program for Jesuit colleges (the so-called *ratio studiorum*) should be contrasted with the fact that one of the greatest opponents of the claim for mathematical certainty, and no less influential than Clavius, was his close Roman colleague Benito Pereira.

Although one might think that the great rationalist philosophers took a clear stand in this battle on the side of the certainty of mathematics, a closer reading of their texts indicates that this view should be handled with care. As soon as one acknowledges that there were different *kinds* of certainty, the picture becomes immediately more complex.

René Descartes, for example, and his frequently quoted methodological treatise, the *Regulae ad directionem ingenii* ("Rules for the Direction of the Mind," written in the 1620s) is a case in point. In the second "Rule," Descartes claimed that: "We should attend only to those objects of which our minds seem capable of having certain and indubitable cognition." Thus, we should reject "all merely probable cognition and resolve to believe only what is perfectly known and incapable of being doubted." This led him to the conclusion that "accordingly, if my reckoning is correct, out of all the sciences so far devised, we are restricted to just arithmetic and geometry if we stick to this Rule." This seems to be, and has often been read as, an unqualified defense of mathematical certainty and as its positing as the only standard for true knowledge. Descartes, like Maurice of Nassau, could certainly claim: "I see that there is nothing certain but mathematics."

However, in the very same treatise Descartes recalled his education in mathematics in the following way:

> When I first applied my mind to the mathematical disciplines, I at once read most of the customary lore which mathematical writers pass on to us. I paid special attention to arithmetic and geometry, for these were said to be the simplest and, as it were, to lead into the rest. *But in neither subject did I come across writers who fully satisfied me.* I read much about numbers which I found to be true once I had gone over the calculations for myself. The writers displayed many geometrical truths before my very eyes, as it were, and derived them by means of logical arguments. *But they did not seem to make it sufficiently clear to my mind why these things should be so and how they were discovered.* (Rule IV, author's emphasis, in Descartes 1985–8: vol. 1, 17)

Descartes further explained that he was therefore not surprised to find "clever and learned men" of his time considering mathematics as a "childish and pointless activity." He went on to say:

> For there is really nothing more futile than devoting our energies to those superficial proofs which are discovered more through chance than method and which have more to do with our eyes and imagination than our intellect; for the outcome of this is that, in a way, we get out of the habit of using our reason. (Ibid.)

Thus, when repeating over and over that mathematics was considered by early modern rationalist thinkers to be a new standard for certainty, one should always be very careful to make precise *what kind* of mathematics and *what kind* of certainty one is talking about. Great categories such as "Platonism" do not help us much here. In the case of Descartes, one of the heroes of our philosophical modernity, there is little doubt that he considered the *existing* mathematical practices as *not* satisfying the standard of the highest certainty.

If he nonetheless maintained that one should take mathematics as a touchstone, it is because he thought that there was a *hidden* ancient knowledge, that should be *uncovered* in order to develop what he called the "true" mathematics (*vera mathesis*). Thus, Descartes is more in accordance with Piccolomini than one would have expected. Reading Euclid, for example, would not provide us with the "most powerful" proofs, to say the least, since the proper mathematical method should make clear to the mind *why* things should be a given way and *how* they were discovered. Accordingly, as we will see again in the next section, Descartes did not much praise the "synthetic" style of geometry practiced by ancient Greek geometers.

Another interesting aspect of the passages mentioned above is to see Descartes referring to "clever and learned men," who saw mathematics as "childish and pointless." This confirms that the discussion about the mathematical standard was not a debate amongst specialists only, but of interest to a large circle of "learned men." But this also underlines that the story about Maurice of Nassau, told at the beginning, stands just for *one* amongst many ways to relate to mathematics amongst these "learned men." In particular, one should not forget that the beginning of the seventeenth century was marked in Europe by a Renaissance of skepticism. This was another motivation for various defenses of the certainty of mathematics (and more generally of science), such as the one written by the Minim priest Marin Mersenne (1588–1648), *La vérité des sciences contre les sceptiques* (The Truth of Science against the Sceptics, 1625). The widespread diffusion of skepticism can also be perceived in Blaise Pascal's (1623–62) reply to "minds excellent in all things else," who were shocked by mathematical paradoxes and used them to criticize their certainty (see below).

It is important to note how Descartes deals with the terms of the debate on certainty: "true mathematics" does not consist of merely stating mathematical facts, but also exhibiting the *why* and the *how*. By the same token, one can detect here a typical feature of early modern views on mathematics: the emphasis put on "invention" ("how they were discovered") and the way in which this approach allows one to unravel how a certain fact occurs. Even if this view was already discreetly present in the "genetic" model provided by Proclus, it acquired a new meaning with the renewal of methodological considerations on the "art of invention" launched by humanist thinkers such as Peter Ramus (1515–72) and the tradition built on his work, which had such a great influence at universities in the vicinity of the Rhine. Moreover, the program of a *nova algebra* (new algebra) presented by Vieta (1540–1603) as a way to uncover the art of mathematical "analysis" (which the ancients were supposed to have hidden from their readers) anchored these considerations in a debate intrinsic to the mathematical practice (cf. Chapter 3). Accordingly, there were two different and often rival strategies for defending the position that mathematics was able to exhibit the "causes." The first emphasized the use of geometrical construction to generate objects with certain properties, whereas the second drew on the new art of symbolic algebra to uncover some constitutive, but hidden relations.

"IN THE MANNER OF THE GEOMETERS"

In relation to the importance of the question of certainty, the early modern period is often seen, more generally, as a time when the very idea of truth underwent a radical transformation. While in ancient traditions truth tended to be understood as knowledge corresponding to a given reality (the standard of *adaequatio*), from the early modern period emphasis would shift to the overwhelming power of the notion of certainty (*certitudo*). In this process, mathematics played a crucial role precisely because it offered a model where inner consistency seemed sufficient to guarantee access to truth. This shift is associated with Descartes and the numerous discussions about his philosophical ideas in the seventeenth century.

Descartes became a celebrated philosopher and mathematician through the publication of his *Discours de la méthode* (Discourse on Method, 1637). The *Discours* served as introduction to three "essays," including his mathematical masterpiece, *La Géométrie*, in which he presented his novel algebraic treatment of geometry (cf. Chapter 3). In the *Discours*, Descartes spelled out four "rules" for the method, the first one being: "to accept nothing as true which I did not clearly recognize to be so," meaning "presented to my mind so clearly and distinctly that I could have no occasion to doubt it." Again, we find certainty in the sense of indubitability as the first standard for truth. This new standard did not only concern isolated ideas. It also pertained to the structure of what true knowledge should consist of. Mathematics, once again, was supposed to provide its only existing model, as a system based on "deduction":

> But now, since we just said that of all the disciplines known by others, only arithmetic and geometry are free from every taint of falsity and uncertainty, we should examine more carefully the reason why this is so. And for this purpose we must observe that we can arrive at knowledge of things by two paths, namely by experience or by deduction. We must observe, further, that while experiences of things are often deceptive, deduction or a pure inference of one thing from another, though it may be passed over if it is not noticed, can never be erroneously executed by an intellect even minimally rational. (René Descartes: *Rules for the Direction of Mind*, Rule II)

It is hence not surprising to see that, when Descartes explained the Rules of his "method" in 1637, he immediately set as a standard "those long chains of reasoning, simple and easy as they are, of which geometricians make use in order to arrive at the most difficult demonstrations" (Descartes 1955: vol. 1, 92).

Many early modern treatises, dealing with all kinds of topics, were written *more geometrico* ("in the manner of the geometers") (Schüling 1969; Arndt 1971). However, to properly assess how the actors related to this "deductive" model, it is very important to keep in mind that many authors praised the standard of mathematics but did not identify this with *existing* mathematics.

Clearly, Descartes, while commending the deductive model of mathematics, did not think that the works of Euclid or Apollonius could stand as its models. As we have seen, he stressed the necessity to show the "why" and the "how." He repeated the same argument in 1641, when asked to present his own philosophy "in the manner of the geometers"—a natural demand considering the praise he gave to the deductive model in the *Discours de la méthode*. In his view, the presentation given in texts such as Euclid's *Elements*, following a "synthetical" order was not suitable to enlighten the mind. Thus, one should prefer an "analytical" order indicating "the true way by which a thing has been discovered methodically" ("Reply to the Second Objections Made to the *Méditations Métaphysiques*," trans. R. Ariew and D. Cress).

Even among authors defending the "synthetic" model inherited from Euclid there were significant differences. The Dutch philosopher Baruch Spinoza (1632–77), for instance, in many respects appears as the quintessential dogmatic philosopher, so deeply instilled with the *more geometrico* standard that he intended to write his complete system of philosophy, his *Ethics,* in this manner. He emphasized that such an enterprise was completely new in its ambition to encompass not only theoretical issues, but ethical ones, and in fact every domain of human existence. In reply to the critique that he attempted "to treat of human vice and folly in the manner of the geometers, and should wish to set forth with rigid reasoning those matters which they cry out against as repugnant to reason, frivolous, absurd, and dreadful," he stressed that the uniformity of the laws of nature commanded the uniformity of the method to understand them (*Ethica*, III, Preface, Spinoza 1883/1884). According to Spinoza, human passions, too, should be treated with the same method, which is employed to know the law of nature, namely "as though [we] were concerned with lines, planes, and solids."

Spinoza is thus a remarkable example of an early modern author who considered mathematics as capable of penetrating every part of human existence. Yet, the geometrical way was thought of as just one way, among others, of presenting a discourse with some advantages *and some drawbacks*. This should be clear from the fact that after having written his *Ethics* "in the manner of the geometers," Spinoza wrote other treatises not presented in this form (like his unfinished *Political Treatise*). Moreover, like Descartes and so many other seventeenth-century philosophers, he had a very specific idea of how mathematics *should* be presented, and this did not correspond exactly to what we find in the ancient Greek treatises. For example, consider what Spinoza wrote regarding the definition of the circle, described in a letter to the philosopher and mathematician Ehrenfried Walther von Tschirnhaus (1651–1708):

> in inquiring into the properties of a circle, I ask, whether from the idea of a circle, that it consists of infinite right angles, I can deduce all its properties. [. . .] If it does not, I look for another, namely, that a circle is the space

> described by a line, of which one point is fixed, and the other movable. As this definition explains the efficient cause, I know that I can deduce from it all the properties of a circle. (Spinoza to Tschirnhaus, January 1675, Spinoza 1883/1884)

A striking feature of this declaration, already present in Descartes, is the emphasis on true mathematical knowledge, which should exhibit the "cause." Another striking feature is that this model of definition, sometimes called "real" or "genetic," is *not* the one which is to be found in the *Elements*. Indeed, the Euclidean definition of the circle typically relies on a property (the fact that some points are situated at an equal distance to a given point, called the "center" of the circle) and not on a genetic process (the moving of the endpoint of a segment, the other extremity of which is fixed).

We find the same emphasis on genetic or "real" definitions in Leibniz. This reference is of particular interest, because Leibniz is often seen as a strong supporter of a different model for the *mos geometricus*, that of "blind" calculation. Indeed, Leibniz claimed on many occasions that if we were able to reduce human thought to some kind of "alphabet" and some basic laws of combination, then every controversy would be reduced to a sheer: "Let us calculate!" (*calculemus!*). His dream was to find symbols or "characters," similar to those of symbolic algebra, but appropriate for the expression of all human thoughts, so that

> all inquiries which depend on reasoning would be performed by the transposition of characters and by a kind of calculus, which would immediately facilitate the discovery of beautiful results. For we should not have to break our heads as much as is necessary today, and yet we should be sure of accomplishing everything the given facts allow. [. . .] And if someone would doubt my results, I should say to him: 'Let us calculate, Sir' and thus by taking to pen and ink, we should soon settle the question. (*La vraie méthode*, 1677, Leibniz 1951: 15)

This program of a *caracteristica universalis* is reminiscent of a very important tradition that spread far into the seventeenth century, and which dreamt of a general "combinatorics" or "calculus" of concepts able to mimic any reasoning and decipher the "secret of the world." This drew on various inspirations, from the Catalan philosopher Ramon Llull (*c.* 1232–1325) and his program of a *Grand Art* to some forms of numerology, inspired by ancient Pythagorism, or by Jewish and Christian Caballa. The *Ars magna sciendi sive combinatoria* (The Great Art of Knowledge, or the Combinatorial Art) published by the Jesuit Anathasius Kircher (1606–80) in 1669 represents a typical example of this Baroque learning (see Figure 4.2). This trend played a crucial role for the development of a mathematical practice of *ars combinatoria* in the seventeenth century.

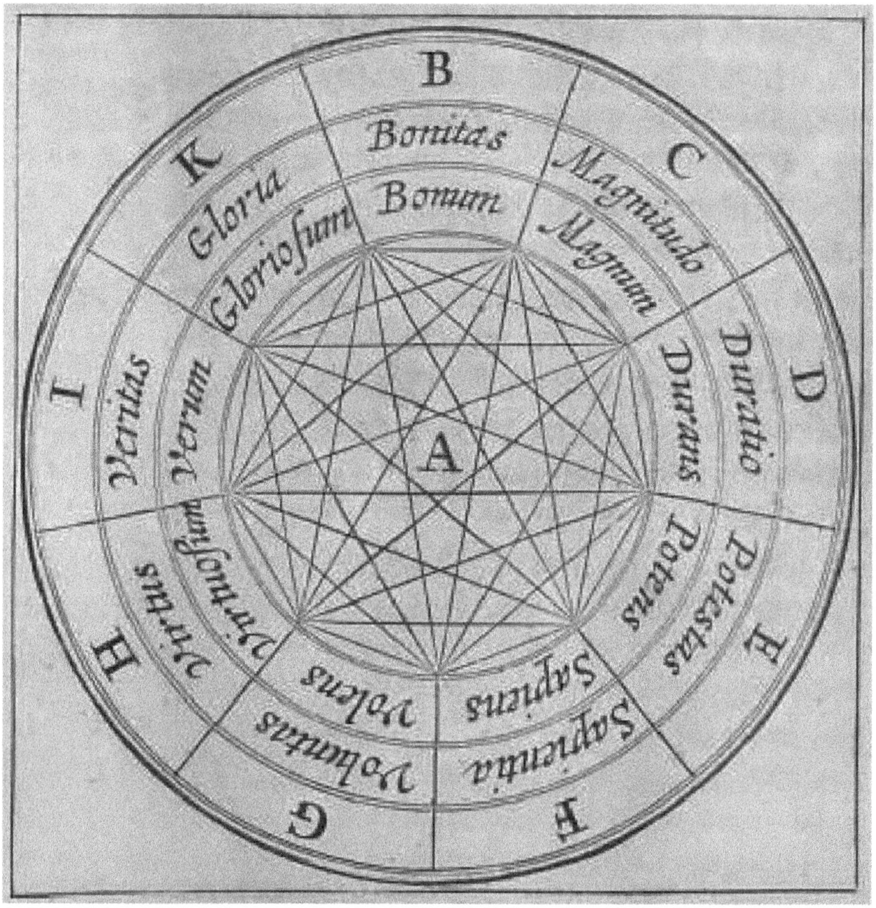

FIGURE 4.2. A Lullian combinatorics of concepts in Athanasius Kircher, *Ars Magna Sciendi, in XII Libros Digesta*, Amsterdam, J. Jansson and E. Weyerstraet (1669: 8). Herzog August Bibliothek Wolfenbüttel: A: 6.3 Quod. 2°.

However, Leibniz, who assuredly drew from this tradition, also expressed on many occasions his doubts regarding our ability to constitute what he called the "alphabet of notions." In absence of a perfect "analysis of notions," he claimed, symbolic computations were a "blind" form of knowledge in the very literal sense that they could contain some hidden contradictions. This pointed to why one should not stop at purely nominal definitions but also provide "real definitions," because the latter indicate not only the meaning of a concept, but its very possibility (i.e., the reason why its object exists).

The three preceding examples, taken from texts written by three monuments of so-called modern rationalism—Descartes, Spinoza, Leibniz—show not only that the mathematical model was object to a large variety of sometimes conflicting interpretations, but also that it was still deeply informed by the terms of the *Quaestio de certitudine mathematicarum*. For all of these philosophers, it was crucial to mathematical knowledge and the methodological model one can derive from it that this approach be capable of exhibiting the "cause" of the properties it deals with.

ON THE INFINITE

According to Alexandre Koyré, during the early modern period, the Universe was, for the first time, identified with infinite geometrical space. As mathematics entered into physics, it destroyed the ancient image of a "closed world" (the Greek "cosmos") and opened up the way to an infinite universe (Koyré 1957). This process was fraught with questions about the relation between the concepts of infinity in mathematics and the traditional ideas discussed until then by the only field legitimized to discuss infinity: theology.

Yet, Koyré's view is very general and does not always concur with the many ways in which philosophers and mathematicians related to infinity. Not only do we find many antecedents to it in Arabic and Indian authors—typically, claims about the infinity of the universe, computations with actual infinite collections, philosophical reflections about mathematical infinity—nor would it be accurate to depict the relationship of early modern thinkers to infinity as detached from the theological framework. Although Descartes, for example, identified matter with geometrical extension and although he famously claimed that all principles of physics were geometrical, he also *denied* any incursion of infinity into mathematics, precisely because it was a realm into which human knowledge had no grasp. In his view, the apparent infinity of the world had the same status as mathematical infinity in the sense that entities were not truly infinite, but "indefinite," i.e. with no apparent limits. As he declared in his *Principles of Philosophy* (1644), "we must not try to dispute about the infinite, but just consider that everything in which we find no limits is indefinite, such as the extension of the world, the divisibility of its parts, the number of the stars, etc." (Part I, prop. 26, Ariew 2000: 237–8). Accordingly, we should not enter into mathematical reflections on the infinite and "never hamper ourselves with disputes about the infinite, since it would be absurd that we who are finite should undertake to determine anything regarding it, and by this means in trying to comprehend it, regard it, so to speak, as finite" (ibid.).

Another interesting case is Blaise Pascal. In one of his most famous "Pensées," called *Disproportion de l'homme* (Disproportion of Man), he intends to describe the "new abyss," in which a human appears to be a "miniature atom" in comparison to the immensity of nature, but also as a "colossus" compared to all the minute entities populating its body and its environment:

> Anyone who considers himself in this way will be terrified at himself, and, seeing his mass as given by nature, supporting him between these two abysses of infinity and nothingness, will tremble at these marvels. I believe that with his curiosity changing into wonder he will be more disposed to contemplate them in silence than investigate them with presumption. (Pascal 1968: 199)

If Pascal has no problem acknowledging an infinite universe, his purpose was clearly not to glorify the way in which the modern thinking subject mastered this infinity. Rather to the contrary, he uses the double infinity as a testimony to the weakness of human knowledge and a reappraisal of Christian faith.

More interestingly, he used the pervasive presence of infinity in mathematics to refute the idea that mathematics could be based on Cartesian certainty. In a small treatise, *De l'esprit géométrique* (Of the Geometrical Spirit), written in the 1650s, he argues that the "perfect method" is inaccessible to human minds since it has to leave some notions undefined and some principles unproven. However, this does not mean that no method is accessible to human knowledge: "For there is one, and it is that of geometry, which is in truth inferior in that it is less convincing, but not in that it is less certain" (trans. W. Wight). Indeed, "it does not define every thing and does not prove every thing, and it is in this that it is inferior; but it assumes nothing but things clear and constant by natural enlightenment" (ibid.). A subtle, but crucial difference to Descartes here is that Pascal does not consider primitive notions, such as space, number or equality, to be "clear and distinct," but "clear and *constant*." If these notions are "constant" (i.e., understood by everybody), and not "distinct," it is precisely because they are related to infinity (of which we have no distinct knowledge): "Thus there are properties common to all these things, the knowledge of which opens the mind to the greatest marvels of nature. The chief of these comprehends the two infinitudes which are combined in every thing: the one of greatness, the other of littleness" (ibid.).

While Descartes thought that the human mind should not enter into the paradoxes of the infinite, because the true infinite was out of its reach, Pascal discussed this matter at length. He did so, however, to reveal that people who spoke of the paradoxes of infinity did not realize how pervasive this notion already was in mathematics. Thus, he intended to clarify the way in which infinity could be *used*, although not *comprehended*. To him mathematics was not a way of putting human beings in a position of mastering the double infinity of the Universe, be it infinitely large or infinitely small, since what is at the core of this knowledge is precisely something *incomprehensible*:

> those who clearly perceive these truths will be able to admire the grandeur and power of nature in this double infinity that surrounds us on all sides, and to learn by this marvellous consideration to know themselves, in regarding

> themselves thus placed between infinitude and a negation of extension, between an infinitude and a negation of number [. . .]. From which we may learn to estimate ourselves at our true value, and to form reflections which will be worth more than all the rest of geometry itself. (*Of the Geometrical Spirit*, Pascal 1910: 444)

In contrast, one might think that the creation of the differential calculus in the 1660s and 1670s (cf. Chapter 3) offers a striking example of how the infinite might be tamed by computations not only in mathematics, but also in our knowledge of nature. Yet, historical studies of the views of the inventors of this new calculus, Leibniz and Newton, have shown that both had very important reservations on this issue. The better known case of Newton is confirmed merely by opening the *Philosophiæ naturalis principia mathematica* (Mathematical Principles of Natural Philosophy, 1687). There, one finds that he did not use his fluxional calculus, but rather a particular version of the geometrical "theory of proportions" that he called the method of first and last ratios (cf. Introduction).

This does not come as a surprise because Newton had serious doubts about the blind use of symbolical computations, which he regarded as closer to heuristics than to proof. Also, he was convinced, like so many of his contemporaries, that geometry was the proper model for certainty (Guicciardini 2009). These doubts, however, were shared by Leibniz, who though a supporter of blind computations was perfectly explicit about the fact that he did not accept infinity in mathematics. Infinity, according to him, should be used in mathematics as something which has only a contextual meaning ("syncategorematic") and does not refer to any kind of actual infinite collections ("categorematic"). It was just an abridged way of speaking and stood for the fact that we can always take some magnitude bigger or smaller than any given one:

> It is perfectly correct to say that there is an infinity of things, i.e. that there are always more of them than can be specified. But it is easy to demonstrate that there is no infinite number, nor any infinite line or other infinite quantity, if these are taken to be genuine wholes. The Scholastics were taking this view, or should have been doing so, when they allowed a syncategorematic infinite, as they called it, and not a categorematic one. (*New Essays on Human Understanding*, II, 17 § 1, Leibniz 1981)

Indeed, for Leibniz, as for Newton, and for many of their contemporaries, theology remained the only legitimate field of inquiry for discussions of genuine infinities, and the great discoveries they made in mastering the infinitesimal calculus had no deep impact on their theological views.

DEBATING THE "NEW MATHEMATICS"

The discussions surrounding differential calculus provide us with an occasion to tackle a larger issue, that of the impact the rapid transformation of mathematics in the sixteenth and seventeenth centuries may have had on worldviews. Indeed, just as infinitesimal methods encountered significant resistance and inertia amongst mathematicians, so were the mathematicians of the preceding period slow to accept the new symbolic algebra completely. While Vieta and Simon Stevin (1548–1620) wrote at the end of the sixteenth century, the algebraization of geometry still had not carried the day until after Descartes published *La Géométrie* in 1637. However, since he wrote it in French and in a particularly (and purposefully!) elusive manner when it came to technical aspects, it did not really further the process of algebraization. Crucial to this development were Frans van Schooten's Latin editions with their additions and commentaries, published between 1649 and 1659–61, thus toward the end of the period dealt with in this volume. It was the latter, far more elaborate edition, that figures such as Leibniz, Tschirnhaus and Newton read and studied. Hence there is an important *caveat* when considering the discussions surrounding the "new mathematics." Although one might have expected a sudden change in the epistemological debates related to important developments in the theory of equations, the transformation of algebra into a symbolic writing, and its application to geometry, the status of algebra was by no means a settled question even during the second half of the seventeenth century.

A striking example is provided by the long-lasting polemics in England between the mathematician John Wallis and the philosopher Thomas Hobbes. Wallis held the prestigious Savilian Chair of Geometry at Oxford and was a prominent supporter of the new algebraic methods. He defended a "Universal Mathematics" (a treatise he published in 1657), in which he expressed the Euclidean "theory of proportions" in a purely arithmetical manner. Hobbes, on his side, was a staunch supporter of ancient constructive geometrical reasoning. He confronted Wallis with the claim that geometry was necessary for a proper understanding of algebra, which he saw more as a language than as knowledge proper (Jesseph 1999). Isaac Barrow, Lucasian Professor of Mathematics in Cambridge, had little sympathy for symbolic mathematics, his geometric style was more in line with Hobbes than with Wallis. Barrow was Newton's teacher and had a long-lasting influence on his deepest views on the nature of mathematical certainty and the pre-eminence of geometry (Guicciardini 2009: 4). Newton represents, in fact, a perfect example of the way a typically "modern" mind can inherit *at the same time* from the development of symbolical algebra (as defended by Wallis), which he mastered like none other, along with the importance of the ancient geometrical standard (as defended by Barrow).

Similar debates played out in France between Antoine Arnauld (1612–94), supporter of the geometric method (although not the ancient one, since he was the author of “new” elements of geometry published in 1667), and Jean Prestet (1648–90), an Oratorian priest and disciple of Nicolas Malebranche (1638–1715) (Mancosu 1996; Schubring 2005). Arnauld raised a question about symbolic algebraic language that was at once simple and deep. If we are to accept that an entity represented by a negative sign is a genuine mathematical object, i.e. a “quantity,” we will indeed reach a strange conclusion. Let us take four quantities A, B, C, D such that A is smaller than C and B is greater than D, it follows from the law of proportionality that the ratio of A to B has to be smaller than the ratio of C to D. Yet, when applying this reasoning to -1 and 1, we get that the ratio of -1 to 1 should be smaller than the ratio of 1 to -1. The discussion between the two scholars occurring during the 1670s was made public with the response Prestet provided at the beginning of the second edition of his *Elemens des mathématiques* (Elements of Mathematics) in 1689. This provides a vivid example of how even late in our period, the issue of the conciliation of modern symbolic techniques and the ancient standard given by the theory of proportions continued to pose tensions.

In 1712, Leibniz recalled the discussions of such questions during his sojourn in Paris (1672–6). He emphasized that this was in fact a deep problem, since the ratio of -1 to 1 is related to the finding of the mean proportional between these two entities (i.e. an x such that -1 is to x as x is to 1), or in other words to a number the square of which is -1. By the same token, this was not only related to “imaginary” roots but also to the logarithm of negative quantities, a subject about which he and Johann Bernoulli (1667–1748) entered into a strong polemic. Be that as it may, the crucial point here is that these debates were typical for the very end of our period and mark a new epoch in the discussions surrounding foundations of mathematics.

STRANGE ENCOUNTERS

The previous sections dealt with the so-called “Scientific Revolution” and the way it brought mathematical “certainty” to the center of “modern” culture. More precisely, important historical nuances have been added to a narrative that contemporary historians increasingly consider a simplifying myth. As we have seen, there is indeed a strong continuity between the emergence of “certainty” as a standard for knowledge, exemplified by mathematics, and previous debates, typical of the Aristotelian medieval university.

But this worldview should also be put into a global perspective. Beyond early modern Europe, a proper tradition of “Japanese mathematics” (*wasan*) began during the Edo period (1603–1867), especially with the work of Seki Kōwa, even though this tradition apparently developed in isolation and did not have a great impact in its own cultural context at the time (Horiuchi 2010). At the same time, there were many ways to transmit knowledge between European mathematicians and, for instance, Arabic or Indian traditions. While the influence of Arabic mathematics had been crucial through its diffusion in Latin medieval treatises and in the practice of abacus schools (cf. Chapter 1), and while many authors had already referred to the "Indian calculus," it is only in the early modern period, that, for the first time, a new and broader *historical* interest in a proper understanding of the development of mathematics in other cultural areas emerged (Stedall 2002).

The debates on the origin of algebra, whether Arabic—as the name suggested—or Greek are a case in point. Rafael Bombelli (1526–72), in the opening to his *Algebra* (1572), relates his surprise on discovering a manuscript of Diophantus in the Vatican Library. But there is more: when reading and translating Diophantus, he came to realize, or so he claims, the importance of Indian mathematics. A similar interest appears, later in the seventeenth century, in the work of John Wallis, who emphasized the importance of Indian mathematics in the birth of algebra (he even presented it as preceding Diophantus) but also in the adoption of a positional concept of number, in which he saw the very foundation of algebra and a testimony of its arithmetical nature (Heeffer 2010).

This period also witnessed the educational activities of the Jesuit order, whose members cultivated, taught and spread the mathematical sciences in Europe and throughout the world. Jesuit missionaries, well trained in mathematics, encountered very rich mathematical traditions, for instance, when they arrived in China (Hsia 2009; Jami 2011). There, the activities of the leading Jesuit personality Matteo Ricci (1552–1610) in cross-cultural transmission of knowledge are highly interesting and provide rich case studies for the confrontation and mutual influences between worldviews. As Joseph Needham once put it:

> In the history of intercourse between civilisations there seems no parallel to the arrival in China in the seventeenth century of a group of Europeans so inspired by religious fervour as were the Jesuits, and, at the same time, so expert in most of those sciences which had developed with the Renaissance and the rise of capitalism. (Needham 1959: 437)

The richest example of these exchanges is provided by astronomical knowledge and the Jesuits' participation in the reform of the Chinese calendar. Since astrological knowledge was at the core of Imperial power in China, this was not just a matter of bringing some light and expelling superstition. A scientific truth could easily turn into a political mistake, as the disgrace of Johann Adam Schall von Bell (1592–1666) shows in a vivid way. After having participated in the creation of the Chongzhen calendar, he rose progressively in the Shunzhi Emperor's court to the point of becoming the head of the *Imperial Observatory*. Yet, after the Emperor's death, he was accused of having willfully made some mistakes in the calculation of certain ceremonial dates. He was then sent to prison, along with some other Jesuits, and died the year after his release, presumably as a result of his imprisonment. It is also important to recall that while Western knowledge was not received without resistance, it also triggered some reflexivity with regard to the history of Chinese mathematics itself, such as can be found in the work of Mei Wending (1633–1721).

To close this narrative, let me end with an anecdote that helps to sustain the picture portrayed above. This concerns the way in which Leibniz reacted to the discovery of some Chinese combinatorial knowledge attached to the practice of the *Yi Jing* (see Figure 4.3). What this ancestral knowledge testified, according to the German philosopher, was yet another instance of the existence of a universal reason (as opposed to the idea that Europe alone was the repository of reason in the world), shared by all cultures and all religions. This kind of universalism may now appear suspect in its blindness to the specificities of other cultures. Moreover, Leibniz was certainly dismissive in his assessment of Chinese mathematics per se: "The Chinese are thus seen to be ignorant of that great light of the mind, the art of demonstration, and they have remained content with a sort of empirical geometry, which our artisans universally possess" (preface to the *Novissima Sinica*, 1697–9, Leibniz 1994).

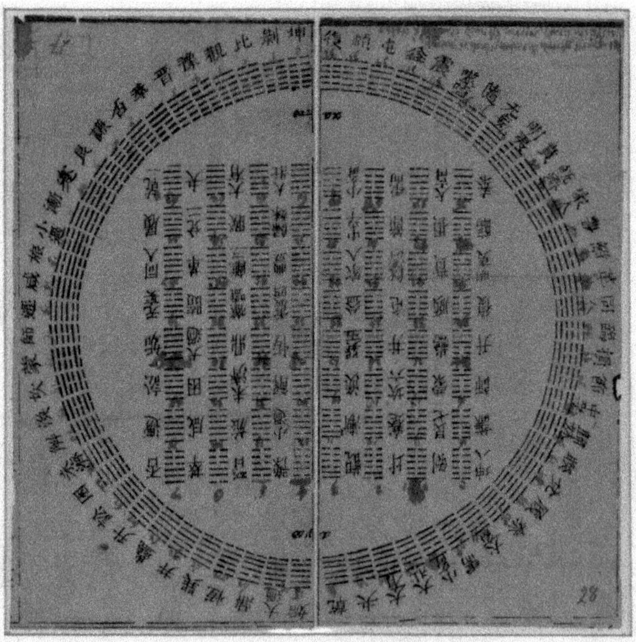

FIGURE 4.3. A diagram for the *Yi-Jing* owned by Leibniz (Letter from Bouvet, 1701, source *Leibniz Archiv*, LBr 105 Bl.27–28). Gottfried Wilhelm Leibniz Bibliothek – Niedersächsische Landesbibliothek, Hannover Signatur: LBr. 105, [53] - 27r.

Yet, it is still worth highlighting that he had no trouble imagining himself as having simply *rediscovered* what was hidden in some ancient Chinese wisdom:

> Fohi, the most ancient prince and philosopher of the Chinese, had understood the origin of things from unity and nothing, i.e., his mysterious figures reveal something of an analogy to Creation, containing the binary arithmetic (and yet hinting at greater things) that I rediscovered after so many thousands of years, where all numbers are written by only two notations, 0 and 1. (*Remarks on Chinese Rites and Religion*, Leibniz 1994)

This is yet another testimony to how complex and rich the history is that underlies the many threads and ramifications between mathematics and worldviews cultivated in the early modern period, but also to the danger of reducing these to oversimplified and homogeneous pictures.

CHAPTER FIVE

Describing and Understanding the World

Setting up the Empire of Mathematics

ANTONI MALET

Early modern mathematics fundamentally contributed to astronomy, optics, mechanics, cosmography, and mapmaking. Mathematical knowledge provided inspiration to design new scientific instruments and many original mathematical results and new curves were invented to describe the world (cf. Chapters 2 and 3). In what follows we will focus on chosen exemplary problems in which mathematics contributed to understanding nature, or in which mathematical inventions answered to natural philosophical questions.

Up to the fifteenth century, the role of mathematics in describing the material world was essentially limited to the science of optics and to Ptolemaic geometrical models that described heavenly bodies' motions. Mathematics also played a role in the medieval science of weights (*scientia de ponderibus*), a discipline that was not central to the medieval curriculum in natural philosophy. As we enter the early modern period, a striking change takes place. Keplerian astronomy was grounded in conic sections and non-uniform motions. New geometrical techniques were introduced in optics, kinematics, and mechanics, sciences that expanded enormously while undergoing radical reformulations. Moreover, new fields and problems appeared that added mathematical methods to their standard tools—mapmaking, artificial perspective, and games of chance being the most substantial ones. These new fields in turn generated new mathematical results.

The chapter is organized thematically. First, it presents the role of mathematics in Copernican and Keplerian astronomy. Then, it turns to the mathematics of cartography, a field that dramatically changed in our period and which posed mathematical problems involving what is now called integration. Next, we turn to optics, which kept then substantial connections with the practice of painting, sculpture, and architecture. Allied with optics, geometry underpinned a major shift in painting, which in itself became a major tool for describing nature. Mathematics also crucially contributed to Kepler's explanation of the optical properties of the eye, and the first theoretical explanations of the properties of optical instruments. We also consider the mathematics of motion and impact, which includes Huygens's interest in horology and his related work on the cycloid. Finally, we give an overview of the first mathematical analyzes of games of chance and their application to actuarial problems. We conclude with a commentary on different approaches used to mathematically describe and explain the world in early modern Europe.

COPERNICAN ASTRONOMY

In 1543, Nicholas Copernicus (1473–1543), a Roman Catholic canon at the chapter of the cathedral of Frauenburg (now Frombork, Poland), published *De revolutionibus orbium coelestium* (On the Revolutions of the Heavenly Spheres). Copernicus's work provided substantial arguments for a heliostatic system of planetary motions, one in which the sun at rest occupies a central position around which orbit the known planets, including the earth (with the moon orbiting it). Copernicus's system challenged Ptolemy's geostatic system, which assumed a motionless earth in central position and provided mathematical models for planetary motions based on combinations of circular motions.

In Ptolemy's models (see Diagram 5.1), planets uniformly turn on a circle (called the epicycle) whose center (P) describes a circle (the deferent) whose center (D) is near the earth (T). P does not rotate uniformly around D, but moves with uniform angular velocity around another point called the equant (E) located on the line joining D to the earth, at the same distance from D as the earth (T) is. Copernicus opposed the use of equants with the argument that P (and the whole epicycle) is not moving with circular uniform speed, thus breaking a fundamental natural philosophical principle concerning the nature of the heavens. Copernicus therefore adjusted his models by adding a second epicycle.

Figure 5.1 reproduces the frontispiece of a major seventeenth-century astronomical work, by the Jesuit Giovanni Battista Riccioli (1598–1671). His *Almagestum novum* (1651) was widely used not only in Jesuit colleges but by professional astronomers across Europe on account of its encyclopedic character, technical proficiency, and updated illustrations and tables. Against

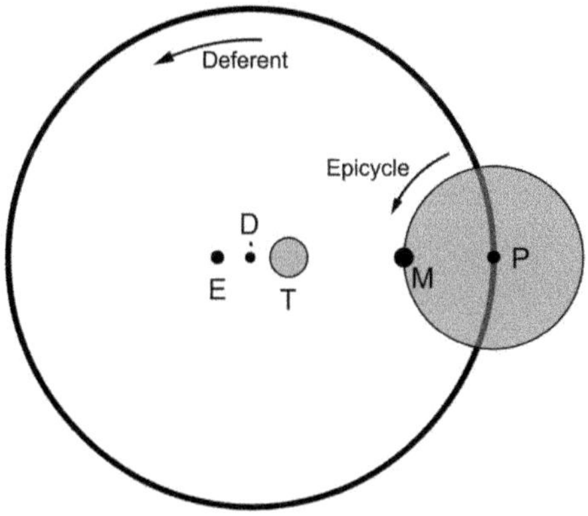

DIAGRAM 5.1. The planet, M, uniformly rotates on its epicycle around center P, which describes a circle (the deferent) centered in D, near the earth T. P moves with uniform angular velocity around the equant, E, located on the line joining D to the earth so that ED = DT. Courtesy of Antoni Malet.

FIGURE 5.1. Giovanni Battista Riccioli, *Almagestum novum astronomiam veterem novamque complectens* (Bologna, 1651), frontispiece with the Copernican and semi-Tychonic systems. Herzog August Bibliothek Wolfenbüttel: A: 2.1.3 Astron. 2° (1).

the Copernican system, Riccioli adopted a variant of the Tychonic system. The Danish aristocrat, Tycho Brahe (1546–1601), amassed for more than two decades a wealth of astronomical observations of unprecedented accuracy, on the basis of which he advocated that the earth stood still at the center of the cosmos (as in Ptolemy's system) while the moon and sun orbited around it. However, the five planets (Mercury, Venus, Mars, Jupiter, and Saturn) orbited around the sun, which carried them around acting as the center of their epicycles. Riccioli and other Jesuit astronomers adopted a modified Tychonic system in which Jupiter and Saturn do not follow the sun but turn around the earth—as shown in the "weightiest" system represented in Figure 5.1. In the frontispiece the Ptolemaic system appears discarded at the bottom right corner.

Copernicus criticized the disconnected, piecemeal nature of Ptolemy's system. Ptolemy put together separate models for each single planet and then added ad hoc (not Copernicus's word) assumptions to account for obvious regularities. The observed constant proximity of Venus and Mercury to the sun, for instance, requires a supplementary hypothesis by Ptolemy—the centers of Venus's and Mercury's epicycles must be aligned with the earth and the sun. Yet, the observed proximity is just a necessary consequence of the Copernican hypothesis. Copernicus used aesthetic values to criticize the Ptolemaic system and its piecemeal nature by recalling the classical story (widely circulated in Humanist circles) about the Greek painter who drew a human body by choosing the most beautiful hands, head, legs, etc., from different models, and ended up with a monster rather than a beautiful figure. The beauty and harmony of a whole is rather more than the sum of beautiful parts. Copernicus also assumed the Neoplatonic and Hermetic trope of the sun as the image of God, which (he said) entitled the sun to occupy the central, commanding position in the cosmos.

Copernicus's work received a mixed reception in the sixteenth century. His heliostatic hypothesis and planetary models were considered intriguing, substantial alternatives to Ptolemaic models, and as such were studied and even applied to calculate new tables. On the other hand, the gains Copernicus's system brought in terms of accuracy in predicting planetary motions were feeble, while the system's downright clash with the physical tenets of the Aristotelian cosmos posed a major obstacle to it becoming a realistic alternative to Ptolemy's system. Things began to change with Johannes Kepler (1571–1630).

Kepler was a convert to Copernicanism from his first contact with astronomy as a Tübingen college student preparing for the Lutheran ministry, which he never attained. Copernicanism fitted well with Kepler's Neoplatonic faith that the cosmos was God's emanation and answered to a geometrical archetype. This led him to the hypothesis that the planetary distances to the sun were determined by the proportional dimensions of the five Platonic, or perfectly regular, solids (tetrahedron, cube, octahedron, dodecahedron, and icosahedron). The sphere of Mercury (the closest planet to the sun) was inscribed in the octahedron, to which the sphere of Venus was circumscribed; then the icosahedron separated the spheres of Venus and the earth; then the dodecahedron determined the radius of the sphere of Mars; then the tetrahedron came before Jupiter, and finally the cube before Saturn. Kepler's first book (1596) demonstrated a rough but acceptable agreement held between the theoretical distances implied by the five solids and the real planetary distances to the sun (as then estimated). He never abandoned his faith that the cosmos was organized according to a divinely inspired geometrical archetype.

With the benefit of Tycho Brahe's observations, which he inherited, Kepler discovered that the shape best describing the motions of planets around the sun were ellipses with the sun occupying one of the two foci. Kepler was using cutting-edge mathematics here, as conic sections had only recently been re-introduced in European mathematics and he was dealing with non-uniform speeds along elliptical paths. Speeds that change from instant to instant posed a major challenge to the mathematical analysis of motion and were a factor that led to the introduction of infinitesimals in seventeenth-century mathematics, as we shall see. Kepler's discovery of elliptical planetary paths may be considered the first historical instance in which conic sections were recognized as necessary to describe an all-important feature of the natural world. Historians of science and historians of ideas have stressed the technical and philosophical implications of abandoning the circle as an archetype for heavenly motions. Alexander Koyré (1892–1964) referred to the episode as marking the end of the 2,000-year-old "haunting spell of circularity" in cosmology.

Kepler's second major breakthrough was his discovery that planetary speeds along elliptical orbits obey the area law (in equal times equal areas are swept out by the line joining the sun and the planet). Finally, he discovered a third major geometrical feature of the planetary system, that for all planets the times required for one full revolution are in the same proportion to the (mean) distances to the sun—namely, the squares of the times are proportional to the distances cubed. The three discoveries have been known as Kepler's laws ever since they played a crucial role in Newton's introduction of universal gravitation. Kepler's laws not only accorded strikingly well with Brahe's observations but they also supported the *Tabulae Rudolphinae* (Rudolphine Tables)—the new astronomical tables Kepler published in 1627 based on his laws and on Brahe's observations—which offered an unprecedented accuracy in predicting planetary positions.

It must be stressed, in conclusion, that mathematics not only provided Copernican astronomy with new curves and new methods to describe planetary motions. Copernicus's and Kepler's philosophical appreciation of geometry's crucial role in the organization of the cosmos substantially contributed to their downplaying the physical arguments against the motion of the earth. The heightened appreciation these authors had for the role of mathematics in astronomy enabled them to challenge the well-established principles of natural philosophy and Aristotelian cosmology. They ushered in an era of intimate connection between advances in astronomy and cutting-edge mathematics.

CARTOGRAPHY IN THE AGE OF DISCOVERY

The extant maps or nautical charts known as portolans, or portolan charts, offer valuable clues on the medieval practice of navigation. They present the most sophisticated and geographically most accurate medieval cartography that we know of, even though they show no grid of coordinates. The most visible and functionally most critical feature of portolan charts is the rhumb lines network. These lines provided a family of basic directions or rhumbs that helped pilots take the best direction to sail along between two given harbors. The basic technique in medieval navigation (dead reckoning) was grounded on direction, as determined by the magnetic compass, together with distance traveled as roughly measured by practical methods. Portolans did not consider the sphericity of the earth, yet since they mostly focused on the Mediterranean basin (a small part of the earth surface approximately falling between the 30 and the 43 degrees of latitude), lines of constant direction could be drawn as straight lines without dramatic errors. Up to the mid-sixteenth century, all nautical charts, including the oldest surviving maps resulting from the exploration of America, answer to the style and technical features of portolan charts.

In the second half of the fifteenth century, when oceanic voyages began in earnest, new methods of navigation were required. In the vast Atlantic, where the determination of position by longitude and latitude became a must, but longitude could not be reliably determined on shipboard on the high seas, Portuguese pilots adopted navigation “by latitude”—once the ship reached the latitude of the island or harbor it was heading to, it sailed due west (or east) until running into its destination. Since ships could be easily thrown off from their chosen parallels, special tables were prepared to help pilots gain or lose one degree of latitude in the quickest way. Three new mathematical instruments answered to the crucial relevance of fixing latitude for fifteenth-century pilots—the quadrant, the mariner’s astrolabe, and the cross-staff, to which the backstaff was added later on. Oceanic navigation required mathematically and astronomically trained pilots. Very early on in the sixteenth century, the crowns of Portugal (in Lisbon) and Castile (in Seville) set up schools to teach pilots plane and spherical geometry, cosmography and astronomy, and navigation “by instruments.” Soon thereafter schools were established in London and elsewhere.

Manuals for the new mathematical methods of oceanic navigation proliferated, some of them going through multiple editions and translations into Latin and the main vernaculars of the time. Perhaps more intriguingly, a new genre of works emerged under the umbrella of “cosmography.” By virtue of their popularity and reputation, Peter Apian’s *Cosmographia* (with commentaries by Gemma Frisius, 1533) and Sebastian Münster's *Cosmographia* (1544) may be taken as paradigms of the genre. In the course of a few decades each of these works went through more than thirty printings and translations in many different languages, including some luxurious editions. Cosmographies offered updated maps along with introductions to astronomy, geography, mapmaking, navigation, theory and use of mathematical instruments, and gnomonics (sundial theory). Their popularity through the sixteenth and seventeenth centuries bears witness to the increasing prestige and status mathematical knowledge was acquiring in early modern Europe.

The first European map including information about the lands newly discovered by Columbus's expeditions was drawn in Cadiz in 1500 by Juan de la Cosa, *c.* 1460–1510, a Castilian pilot. It is inked on vellum and looks like a portolan chart—as suggested by style, ornamentation, and above all by what has the appearance of the typical network of straight lines representing rhumb lines in portolan charts. As explained above, portolan rhumb lines roughly corresponded (within the margin of error of contemporary nautical knowledge) to Mediterranean trajectories travelled following a set rhumb, or geographical direction. Furthermore, it was assumed that by sailing according to the rhumb determined by any straight line on the chart, the ship took the shortest path—that is to say, on the assumption of a flat sea surface, straight lines on the chart corresponded to straight lines between harbors. These assumptions crumbled with the experience of transatlantic navigation.

The great Portuguese mathematician Pedro Nunes (1502–78) first clarified the mathematical nature of trajectories of constant bearing on a spherical surface (rhumb lines) by distinguishing them from the shortest path between two points (a great circle). Nunes demonstrated that trajectories along a fixed bearing (other than sailing due east/west or due north/south) described a spiral line around the pole which draws ever closer without ever reaching it. In 1624, the Dutch mathematician Willebrord Snell (1580–1626) called rhumb lines on a sphere *loxodromes*, a name still used today. More generally, Nunes explored some of the geometrical problems involved in making nautical charts, or "mapping" the spherical earth upon a plane. Nunes's foundational work on mathematical navigation and cartography was influential. It appeared first in Portuguese in 1537 and then in a longer and more sophisticated version in Latin in his *Opera* (1566). He corresponded with and inspired John Dee (1527–1608) in London and was closely read in the Netherlands (then ruled by the Hispanic monarchy), the base of a large community of Portuguese Jews (Nunes was a converted Jew) that was to become the capital of European cartography in the mid-sixteenth century.

In 1537, along with his pioneering tracts on loxodromes and nautical charts, Pedro Nunes edited Book I of Ptolemy's *Geography*, which discussed map projections. Ptolemy, who listed terrestrial positions with their geographical coordinates, criticized the projection that mapped the parallels and meridians of the earth's surface into a flat grid of perpendicular straight lines by pointing out that spherical parallels are represented by the same length, regardless of latitude. He suggested two alternative projections for drawing a map of what was then considered the inhabited world on a plane surface.

Latin translations of Ptolemy's *Geography* appeared early in the fifteenth century, to be widely used in teaching geography at universities. By the end of the fifteenth century, this text was instrumental in recovering the idea of longitude and latitude grids in mapmaking. Cartographers introduced modifications to Ptolemy's projections and designed new ones, some of them working in Leuven in Peter Apian's and Gemma Frisius's circles. A student of Frisius, Gerard Mercator (Geert de Kremer, 1512–94), introduced one of the most innovative ideas in early modern mathematical cartography.

What is commonly called the Mercator projection was first used in 1569 to produce the large world map Mercator published as *Nova et aucta orbis terrae descriptio ad usum navigantium emendate accomodata* (New and More Complete Representation of the Terrestrial Globe Properly Adapted for Use in Navigation, 124 × 202 cm, Duisburg, 1569). It is not known how Mercator plotted his map, but it may be described as a cylindrical projection—from the sphere's center, every point on the surface is projected onto the surface of the cylinder tangent along the equator (in theory, the poles go to infinity, but for latitudes above 60°N and S, a correction was applied). The cylinder is then cut open along a line perpendicular to the equator and unrolled into a plane map. Meridians are thus mapped into equidistant vertical straight lines while parallels turn into horizontal straight lines. As the latitudes increase, on the same meridian equal differences in latitude are mapped ever longer (5° of latitude near the equator are mapped into a much shorter line than 5° at the latitude of Paris, *c.* 50°N). Mercator never explained the mathematics he used to plot parallels and positions on his map. The only indication he gave was that his projection lengthens the degrees on the meridians in the same proportion as the parallel degrees do. He rightly assumed that angles would thereby be preserved and loxodromes would be represented as straight lines, which was in fact the case in theory and on his map, although no mathematical demonstration was produced until the 1690s. This was all Mercator said about his new map.

The mathematical grounds of Mercator's idea were publicly known only when the London based mathematician and instrument maker, Edward Wright (1561–1615), published *Certaine Errors of Navigation* (1599). Acknowledging Pedro Nunes's pioneering efforts, Wright pointed out that if two given meridians were separated at the equator by an arc of length λ, the length of the arc separating them on the parallel at latitude θ would be $\lambda\cos\theta$. Therefore, if the parallels are projected into a grid of perpendicular straight parallels and meridians, they lengthen by a factor $1/\cos\theta$ (or $\sec\theta$). Accordingly, Mercator's projection required the degrees of mapped meridians continuously to grow with the latitude θ by the factor secθ. Wright solved the problem by numerical approximation. He divided the meridian in parts of 1' and computed the length of the mapped meridians by adding the equatorial length of 1' (one minute = one sixtieth of 1°) times the secant for all the successive latitudes (namely, starting from 0 and adding for angles increasing by 1' up to 75°). With that he produced a 23-page table (published in 1610) with the lengths of "Wright's meridian parts," as the results became known in seventeenth-century England. The modern reader familiar with the ideas and symbols of the calculus (which became available only after 1690) may have recognized that on the map the length of the meridian from the equator to the latitude θ is $\int_0^\theta \sec\theta\, d\theta$, and that Wright performed a numerical integration *avant la lettre*.

In the second quarter of the seventeenth century, as numerical tables of trigonometric ratios and their logarithms proliferated, it was only a matter of time before practitioners began to remark upon the striking similarity between Wright's table for "meridian parts" and tables for the logarithms of tangents. By the mid-century, Richard Norwood (1590?–1675) pointed this out in print. In 1666, the London-based mathematician, Nicolaus Mercator (1619–87, no relation to Gerard), posed the question as an open problem in the *Philosophical Transactions*, the widely read, unofficial journal of the Royal Society: "Whether the Artificial [logarithmic] Tangent-line be the true Meridian-line, yea or no?" The answer is yes, but the demonstration was in pre-calculus times a difficult problem in higher mathematics. It was solved and published by James Gregory (1638–75) in 1668.

Nautical loxodromes were related to a curve newly introduced in the seventeenth century (it first appeared in a dynamical problem) and called the logarithmic spiral. This plane curve meets all the radii coming from a fixed center, O, under a constant angle, revolving infinitely often as it spirals around O. In the logarithmic spiral, as the angles increase in arithmetical proportion, the radii decrease in geometrical proportion, whence its name. By the 1660s, at the latest, mathematical practitioners working on navigation and mapmaking hoped to find methods for rectifying (finding the length of) arcs of the logarithmic spiral. These might then prove useful for determining the rhumb lines between two points whose latitudes and difference in longitude were known, which could also perhaps contribute to finding longitudes. In these terms and in the years around 1670, John Collins (1625–83)—Fellow of the Royal Society (FRS) and unofficial mathematical assistant to the Royal Society's Secretary—posed this difficult problem (and others of a practical nature) to John Wallis, Isaac Newton, James Gregory, and others. Only Gregory worked out a solution, which led him to first find the quadrature of the logarithmic curve (a major problem in itself). Intriguingly, the Royal Society did nothing with Gregory's results, which were rediscovered among his manuscripts in the twentieth century—perhaps they were too sophisticated for practitioners' needs.

The logarithmic spiral was taken up by Jakob Bernoulli (1655–1705) in the early 1690s. He mentioned the spiral's historical links with nautical and cartographical problems, but his approach was theoretical and abstract. In 1696, Edmond Halley (1656–1742) reviewed most of the foregoing developments in the *Philosophical Transactions* (n° 219: 202–14), including his improved techniques for the numerical computations associated with the quadrature of the secants. He concluded by stressing that the numerical methods already available were "abundantly sufficient for nautical uses" and that his own contributions answered a theoretical rather than an actual practical requirement, since in fact his new method provided accuracy for which there was no occasion.

Cartography answered to the need of all political powers across early modern Europe for the production of accurate representations of the lands and seas of the earth. Closely connected to the new "cosmographies" and to mathematical navigation, cartography was deeply dependent on mathematics, while inspiring the invention of new instruments. It contributed to increasing the value and prestige of mathematics and expanding mathematical education, which occupied pride of place for many new professions, including pilots, mapmakers, and instrument makers. Interestingly, cartography and navigation posed a number of practical problems that directly led to deeper and more abstract mathematical developments, whose solutions turned out to have no practical use.

NATURAL AND ARTIFICIAL PERSPECTIVE

Optics was taught in medieval universities, where it was known by the Latin name *perspectiva*. It was a well-organized discipline in which geometry played a major role. "Artificial" or pictorial perspective took shape on the theoretical grounds of medieval tradition. *Perspectiva* was often read in the thirteenth-century synthesis written by Witelo (or Vitello), which was heavily indebted to the Latin translation of the treatise *Kitab al Manazir* (Book of Optics) by the eleventh-century Arabic author Ibn al-Haytham (Alhazen in the Latin West). Both Witelo's *Optica* and the Latin version of Ibn al-Haytham's *Book of Optics* circulated widely in sixteenth-century Europe. As suggested by its title, Johannes Kepler's *Ad Vitellionem paralipomena* (Complements to Witelo, 1604) was modestly presented as a commentary on Witelo's work. In fact, however, Kepler produced a major geometrical reformulation of the principles of optics, including the optical workings of the eye's transparent humors.

In the *perspectiva* tradition, vision was accomplished by the visual "pyramid" made up of the visual rays issuing from every visible point on the observed object and converging towards the eye, in whose center the vertex lies. Only rays impinging perpendicularly on the external surface of the eye were assumed to produce vision—the neglected rays were supposed to be ineffectual because their inclined incidence made them defective

The perpendicularity condition establishes a 1-to-1 mapping of visible points before the eye onto points on the cornea. Beyond the cornea, the rays were assumed to meet the crystalline humor in the same arrangement as when they moved towards the eye. The crystalline humor or lens was assumed to be the organ of vision that enabled the rays to become visible to the eye. Henceforward vision moved to the optical nerve and then on to the brain. The foregoing understanding of vision must have been widely shared, as it was taught with minor differences in colleges and universities across Europe and is found in many perspective tracts meant for use in workshops by painters and architects up to 1600 and beyond. The visual pyramid has been recognized as a major influence leading to "artificial" or linear perspective.

In the fifteenth and sixteenth centuries, painters paid heed to what optics teaches about perception. In a then common metaphor, a painting should be a "window" opening on the world outside. Accordingly, the painter sought to represent the three-dimensional "outside" on the flat surface of the painting in such a way that this flatness disappears to create instead the illusion of depth and three-dimensional space. Taking their cue from optics, painters conceptualized the painting surface as a geometrical plane intersecting the visual pyramid reaching the eye of the beholder at a fixed point (cf. Figure 5.2, taken from one major seventeenth-century reference on perspective, *Le due regole della prospettiva pratica* (Rome, 1583), by Jacopo Barozzi da Vignola (1507–73), edited posthumously by the mathematician Egnatio Danti).

The oldest extant treatise on geometrical perspective, *De prospettiva pingendi*, was written *c.* 1480 or shortly before by Piero della Francesca (*c.* 1420–92), a quattrocento master who was also a distinguished mathematician. Piero assumed *transversals* (lines parallel to the picture plane lying on the floor or on a plane parallel to the floor) should be represented on the painting as lines parallel to the ground line, and that *orthogonals* (lines perpendicular to the picture plane) should

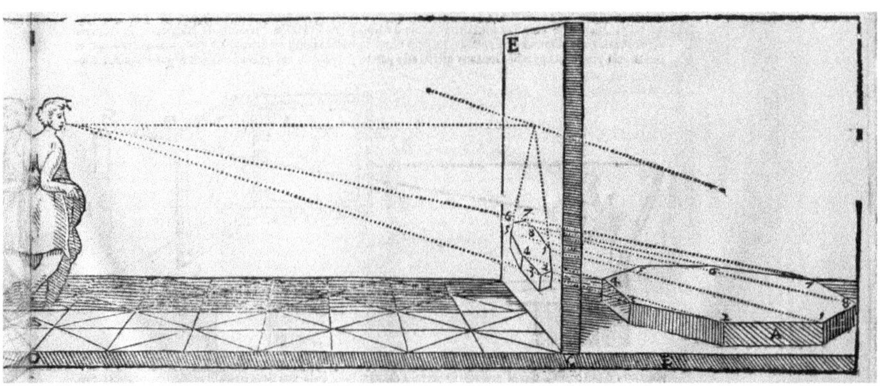

FIGURE 5.2. The painting plane intersecting the visual pyramid in Vignola's *Le due regole della prospettiva pratica* (p. 55). © Universitätsbibliothek Heidelberg.

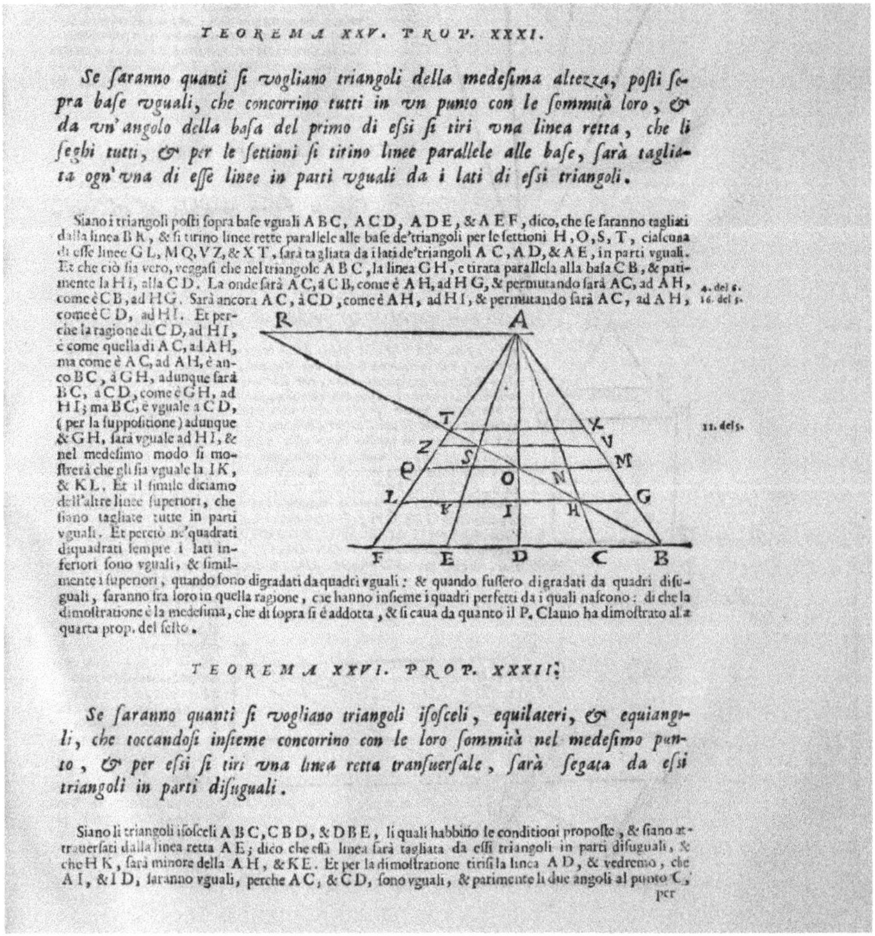

TEOREMA XXV. PROP. XXXI.

Se saranno quanti si vogliano triangoli della medesima altezza, posti sopra base vguali, che concorrino tutti in vn punto con le sommità loro, & da vn'angolo della basa del primo di essi si tiri vna linea retta, che li seghi tutti, & per le settioni si tirino linee parallele alle base, sarà tagliata ogn'vna di esse linee in parti vguali da i lati di essi triangoli.

Siano i triangoli posti sopra base vguali A B C, A C D, A D E, & A E F, dico, che se saranno tagliati dalla linea B R, & si tirino linee rette parallele alle base de'triangoli per le settioni H, O, S, T, ciascuna di esse linee G L, M Q, V Z, & X T, sarà tagliata da i lati de'triangoli A C, A D, & A E, in parti vguali. Et che ciò sia vero, veggasi che nel triangolo A B C, la linea G H, e tirata parallela alla basa C B, & parimente la H I, alla C D. La onde sarà A C, à C B, come è A H, ad H G, & permutando sarà AC, ad A H, come è C B, ad H G. Sarà ancora A C, à C D, come è A H, ad H I, & permutando sarà A C, ad A H, come è C D, ad H I. Et perche la ragione di C D, ad H I, è come quella di A C, ad A H, ma come è A C, ad A H, è anco B C, à G H, adunque sarà B C, à C D, come è G H, ad H I; ma B C, è vguale à C D, (per la suppositione) adunque & G H, sarà vguale ad H I, & nel medesimo modo si mostrerà che gli sia vguale la I K, & K L. Et il simile diciamo dell'altre linee superiori, che siano tagliate tutte in parti vguali. Et perciò ne'quadrati diquadrati sempre i lati inferiori sono vguali, & similmente i superiori, quando sono digradati da quadri vguali: & quando fussero digradati da quadri disuguali, saranno fra loro in quella ragione, che hanno insieme i quadri perfetti da i quali nascono: di che la dimostratione è la medesima, che di sopra si è addotta, & si caua da quanto il P. Clauio ha dimostrato al a quarta prop. del sesto.

4. del 6.
16. del 5.
11. del 5.

TEOREMA XXVI. PROP. XXXII.

Se saranno quanti si vogliano triangoli isosceli, equilateri, & equiangoli, che toccandosi insieme concorrino con le loro sommità nel medesimo punto, & per essi si tiri vna linea retta transuersale, sarà segata da essi triangoli in parti disuguali.

Siano li triangoli isosceli A B C, C B D, & D B E, li quali habbino le conditioni proposte, & siano attrauersati dalla linea retta A E; dico che essa linea sarà tagliata da essi triangoli in parti disuguali, & che H K, sarà minore della A H, & K E. Et per la dimostratione tirisi la linea A D, & vedremo, che A I, & I D, saranno vguali, perche A C, & C D, sono vguali, & parimente li due angoli al punto C,

per

FIGURE 5.3. A simplified method to draw the image of a chequered square lying on the floor, from *Vignola's Le due regole della prospettiva pratica* (p. 37). Max Planck Institute for the History of Science, Berlin.

be represented as converging towards a point determined by the perpendicular projection of the eye upon the painting, the so-called main vanishing point (point A in Figure 5.3). These assumptions were consequences of the optical framework just mentioned, and in particular of the geometry of the visual pyramid. Assuming the eye's height above the floor and its distance to the picture plane were known, Piero produced a number of geometrical methods to project geometrical objects of varying complexity upon the picture plane. The most basic result concerned the projection of a checkered square or rectangle, a projection that became a hallmark of High Renaissance paintings. The main vanishing point on the painting, A (cf. Figure 5.3) is given, and the distance from the eye to the picture plane is also known. FE = ED = . . . are the sides of the floor squares. The lines joining F, E, . . . with A are the images of the orthogonals. AR is parallel to the ground line, FB, and the distance AR is equal to the distance from the eye to the picture plane. Then Piero drew the straight-line RB, whose intersections T, S, O, H with FE, EA etc. determine the transversals TX, ZV, QM, LG. By using similar triangles, perspective writers demonstrated this method to yield the same result as the direct projection shown on Figure 5.2.

After 1500, geometrical tracts proliferated that catered to the needs of an ever-growing number of artists and artisans involved in one way or another in image production—including stage designers, sculptors and goldsmiths, woodcarvers, engravers, and stone-cutters, not to mention engineers, architects, and painters. Tracts were often written by artists and artisans hoping that authorship would help them attain higher status. This is also evidence of the rising status of mathematics itself (cf. Chapter 4). Many tracts merely taught geometrical recipes for the easy construction of images "in perspective." However, a number of distinguished mathematicians, including Federico Commandino (1509–75), Guidobaldo del Monte (1545–1607), Simon Stevin (1548–1620), and Girard Desargues (1591–1661), devoted important works to investigate purely geometrical questions arising in linear perspective. Risking oversimplification, in the hands of creative mathematicians, geometrical perspective was liberated from optical assumptions and preoccupations to become a fully grounded mathematical discipline, based on a general theory of central projection upon a plane. The contributions of Guidobaldo and Stevin in particular provided inspiration for Desargues' early work on projective geometry.

A noteworthy application of the geometry of perspective to painting is found in anamorphoses. The result of a distorted projection, anamorphoses produce a normal appearance only when viewed from a particular point. Perhaps the most well-known of early modern anamorphoses is the skull that breaks the symmetry of the floor tiles in Hans Holbein's masterpiece, *The Ambassadors* (1533). Only when the beholder's eye is located some two meters off to the right side of the painting, about the height of its upper side and close to the picture plane, does the skull appear with all its grim evocative power. The visual tension between a perfect geometrical projection on the painting surface that cannot be "read" from the ordinary point of view, but requires a viewpoint wherefrom the rest of the painting is blurred, was put to the service of moral and religious messages. Anamorphoses were a favorite of mathematicians in Roman Catholic orders. The foremost textbook on their theory and practice, *Thaumaturgus opticus* (1646), was penned by the Minim father J. F. Niceron (1613–46), a student of Marin Mersenne. Anamorphoses graced the walls of many religious houses, most of them now lost, but one magnificent anamorphic fresco—designed in 1647 by another prominent Minim mathematician, Emmanuel Maignan—is still extant in the Trinità dei Monti church in Rome (Andersen 2006).

KEPLER'S EYE AND GEOMETRICAL OPTICS

Mathematics played a crucial role in Kepler's elucidation of how the eye works, a major breakthrough in seventeenth-century natural philosophy. Kepler's optics of the eye occupies pride of place in his *Complements to Witelo* (1604) and led him to rethink the nature of optical images, which eventually led to his theory of the telescope in his *Dioptrice* (1611).

Kepler argued that the eye was best understood in terms of a camera obscura instrument. In a camera obscura, or dark chamber, a very thin opening (or pinhole) lets light in to be projected upon a white screen or wall, upon which a symmetrical inverted image appears of whatever is outside and facing the opening. Sixteenth-century dark chambers were courtly attractions as well as observational aids for astronomers, particularly for observing solar eclipses.

The classical anatomy of the eye accurately described the eye's three main sections or chambers—the aqueous humor in the anterior cavity, the crystalline humor (or lens) in central position, and the vitreous humor behind the lens. In his *Complements*, Kepler mentioned an informal suggestion made by the famous polymath, Giovanni Battista della Porta (1535–1615), in his European bestseller *Magia naturalis* (1589). The pupil, Della Porta said, acts as a camera obscura pinhole to let "the likeness" (*idola*) of external objects in to have them projected upon the anterior, more external surface of the crystalline humor. Della Porta's brief passing reference, devoid of any argument or detail, did not intend to question the tenets of medieval optics. By default, Della Porta's reader is led to believe that the crystalline lens, instead of receiving the *species* of the visual pyramid directly, as an intersection of the pyramid, receives them symmetrically inverted by the pinhole effect.

In Kepler's theory of the eye, the retina—not the crystalline humor—plays the role of a screen upon which the outside world is symmetrically projected. But this is a deceptively simple way of comparing Kepler's theory with Della Porta's suggestion. Kepler first revealed the geometry that explains the formation of pinhole images, a question that had beset optical authors since antiquity. He showed pinhole symmetrical images to be simply necessary consequences of light propagating in straight line rays through very small openings. Kepler thus questioned the notions of *species* and *idola*—he simply let light rays do all the work.

Next, Kepler turned to understand what happened to light as it entered the eye and was refracted by transparent humors. (Snell's law of refraction—already published by Ibn Sahl in 984—came only later, but Kepler used a good approximation, and for angles $\leq 30°$ made the angle of refraction proportional to the angle of incidence.) In particular, given the prominent role of the crystalline humor in the then prevalent theory of vision, Kepler focused on it. We call the crystalline humor the *lens* of the eye, but before 1600 no one understood how lenses work—although they had been in use since the fourteenth century. In order to unveil the optical properties of the crystalline

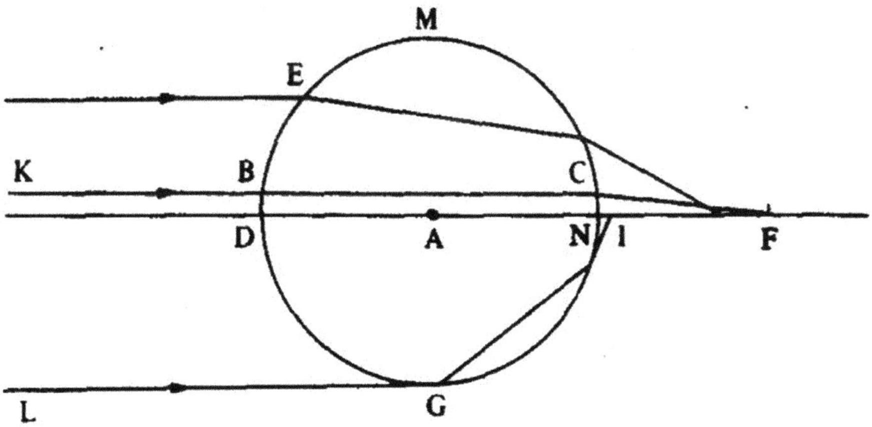

DIAGRAM 5.2. From Kepler's *Paralipomena* (p. 188), simplified. Courtesy of Antoni Malet.

humor, Kepler assumed it produced optical effects like those produced by a spherical vial full of water, so he analyzed the geometry of refractions through such a spherical vial (cf. Diagram 5.2).

Parallel light rays falling upon a transparent spherical vessel filled with water (with center A) are twice refracted, first as they pass into the vessel and then as they emerge out of it—as in E and C for a particular ray. Kepler demonstrates that the farther any incoming ray is from the diameter (like LG), the nearer it meets the prolongation of the diameter (at I) as the ray goes out of the vessel. Conversely, the closer the rays are to the diameter, the farther away they meet it after coming out of the vessel (as KBF). He also demonstrated that the radius length, AN, is an upper bound for the distances at which refracted rays meet the prolongation of diameter DN.

By placing a spherical flask full of water up against the pinhole of a camera obscura, Kepler obtained a "nice distinct" image of the outside objects drawn on a white paper placed at a distance equal to the vessel's radius. Moving the vessel either closer to or farther away from that location, the image became blurred. Kepler's analysis of the geometry of refractions accounted for this, as he was led to recognize that most rays passing through the water parallel to the diameter DAN meet the prolongation of the diameter very close to the point F (cf. Diagram 5.2), whose distance to the vessel is equal to the radius AN. He was thus led to the idea of a focal point for refracted rays by relying on an informal, *avant la lettre* conceptualization of a geometrical limit—a major feat.

He completed his analysis by looking at the refraction of rays coming from points outside the axis, such as I (cf. Diagram 5.3), when a diaphragm (EF) reduces the radiation falling on the spherical vessel. In this situation—which Kepler explicitly meant to be a geometrical model for the pupil and crystalline humor of the eye—he demonstrated two things. First, he showed that the image

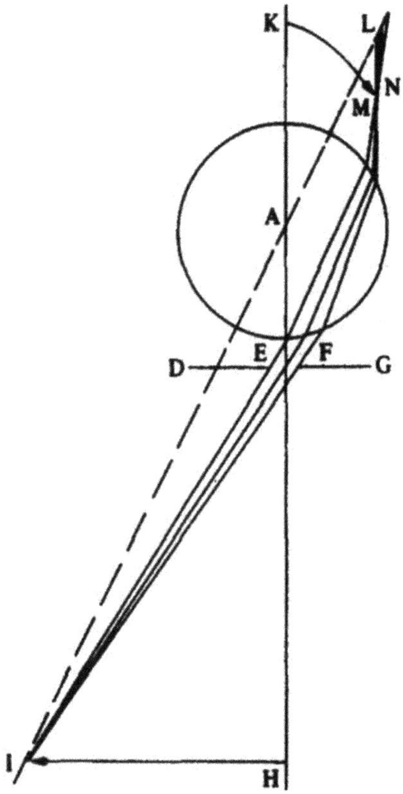

DIAGRAM 5.3. From Kepler's *Paralipomena* (p. 196), arrows added. Courtesy of Antoni Malet.

of H (at K) would be very distinct because the diaphragm only allows rays very close to the diameter to be refracted, and therefore the focus at K is most sharply defined. Second, that rays coming from I would not be refracted along the axis HA, but as indicated in the figure would have their focal zone at MN, and therefore the image of IH would appear most distinctly on a curved surface, rather than on KL. To Kepler this provided a teleological explanation for the quasi-spherical shape of the retina.

Kepler stressed that the distinct images "painted" (as he put it) on the retina had special qualities when compared with those on mirrors or with the image of a rod half immersed in water that looks as if it bends at the water surface. The latter images, said Kepler, are deceptions—they make the viewer think objects are out of place or deformed, unlike those drawn on the retina, which are different in kind. Images drawn on the retina materialize on a screen-like surface; their shape and the spatial location of the screen are geometrically determined. They are made up of light rays falling on the screen; and (leaving aside the symmetrical inversion) they are most true to nature. Kepler called them *picturae* to stress their faithfulness and materiality, using the word *imago* for the others (images on mirrors and so on). Kepler left it to the "physicians" (*sic*) to explain how the geometrically determined *picturae* were perceived by the spiritual human soul, but he suggested that the geometrical essence of *picturae* made them directly apprehensible by the soul.

Kepler put these results to use in analyzing the optical effects of convex and concave lenses, to which he applied the notion of *pictura* to explain the production of what we now call "real" (as opposed to virtual) geometrical optical images. In 1610, Galileo sent Kepler a copy of his *Starry Messenger*, the little book containing Galileo's sensational and controversial telescopic discoveries. In 1611, Kepler published *Dioptrice*, a little tract with a geometrical theory explaining the Dutch (or Galilean) telescope, in which a convex objective combines with a concave ocular to magnify vision. *Dioptrice* provided a general theoretical framework for optical instruments in the first two thirds of the seventeenth century.

Early modern optics, painting, and geometry mingled together, an intimate union forged, so to speak, by the fifteenth-century turn in Western culture towards an understanding of art in which *imitatio* (direct imitation of reality) was a major concern. Geometry became a bridge of sorts between optical assumptions and painters' training, while geometrical knowledge drew inspiration from problems born in linear perspective. Geometry was essential in shaping and grounding Kepler's new theory of the eye (the same may be said for his new understanding of optical images), whose immediate contemporary success across the philosophical spectrum (from Descartes to the Jesuits) owes much to its consistent geometrical scaffolding.

THE MATHEMATICS OF MOTION

The deep transformation of gunfire and the rapidly increasing relevance of cannons in warfare made tracts on the art of gunnery highly popular in the sixteenth century. These works included problems in ballistics, such as the determination of the projectile's path, how to produce the farthest shot, or the most violent shot, and so on. A very able mathematician, Niccolò Tartaglia (1500–57), wrote one of the earliest and most popular sixteenth-century treatises on gunnery—which he simply entitled *Scientia nova* (New Science, 1537). It is written in classical geometrical style, starting with definitions and four axioms that express traditional dynamical ideas

In Tartaglia's analysis, the projectile first follows the motion imparted to it by the cannon shot. This motion dominates the first stage of the cannon ball's trajectory, producing a straight line up until the second stage, which is shaped as a circular arc. The third stage of its trajectory commences with the ball's vertical fall downward under its natural tendency as a heavy object. Tartaglia provided a wealth of geometrical diagrams to represent the trajectory, but he did not properly explain what happened during the second, circularly shaped stage—a puzzling problem, given his explicit claim that impressed and natural motions are incompatible. In his *Quesiti et inventioni diverse* (On Various Questions and Inventions, 1546), Tartaglia changed his position and argued that, excepting vertical shots, trajectories are completely curvilinear, though he did not elaborate on the idea.

Tartaglia's first theorem proves that the speed of free-falling bodies increases with the distance from the starting point—which makes speed proportional to the space traversed, a popular hypothesis at the time. With hindsight, Tartaglia's book shows the difficulties inherent in trying to cast the principles of mechanics in geometrical terms. Other attempts were made in sixteenth-century Italy to set up a geometrically consistent framework to describe objects in motion, among which perhaps the most philosophically sophisticated were those of Giovanni Battista Benedetti (1530–90). The task was essentially accomplished by Galileo Galilei (1564–1642), though because of the mathematics available to him, Galileo's success in grounding the mathematics of motion was harder to achieve than it may appear. In sixteenth-century pure or scholarly mathematics (as set forth in Euclid's *Elements*, but in opposition to what was accepted in practical geometry and arithmetic), measurements of geometrical magnitudes (lines, surfaces, and so on) were never used, which meant that no arithmetical operations were defined among them. Consequently, one could compare magnitudes of the same kind—lengths with lengths, times with times, speeds with speeds, etc., but lengths could not be compared with times, for instance. Ratios between magnitudes were handled by the convoluted Euclidean definition of equality of ratios in Book V of the *Elements*. For Galileo, as for any early modern natural philosopher, speed was a *primary* property of bodies in motion, not a concept *defined* as a ratio of space to time—a definition not introduced until *c.* 1700.

Galileo's principles of motion were published in 1638 in a treatise entitled *Discourses and Mathematical Demonstrations Relating to Two New Sciences*. He stated that, for bodies moving uniformly, the longer the time of motion, the longer the distance traversed; and conversely, the longer the distance, the longer the time. In general, for a given interval of time, the greater the speed, the longer the distance traversed; and conversely. We get a sense of Galileo's conceptual framework by looking at the first *theorem* of his science of motion: "If a particle, moving uniformly at a constant speed, traverses two distances, the time-intervals required are to each other as the ratio of these distances," or with anachronistic notation, if the speed is constant, then $t_1 : t_2$ equals $d_1 : d_2$. Notice that, translated into our idiom, constant speed means d / t does not change, and there is nothing to prove. However, Galileo and his contemporaries could not define speed as the ratio of distance to time, because distances and times could not have a ratio. Galileo thus had to prove this theorem by applying Euclid's cumbersome definition of equality of ratios.

Galileo dodged discussions ensnaring his predecessors by avoiding the issue of causality, about which he only remarked: "the present does not seem to be the proper time to investigate the cause of the acceleration of natural motion" (Galilei [1638] 1954: 166). Instead, he focused on geometrical description. Naturally accelerated motion (Galileo's next topic) required handling speeds that change from one instant of time to the next. Since mathematics did not yet have a proper place for the infinitely small, non-uniform speeds posed a major conceptual difficulty. Galileo defined uniformly accelerated motion as that in which its moments of swiftness (*celeritatis momenta*) "receive equal increments in equal times," but he left the notion of "moment" undefined.

Galileo's first main result on the topic of uniformly accelerated motion was his proof of the famous mean-speed theorem (cf. Introduction to Volume 2). This states that the distance traversed by a body starting from rest with uniformly accelerated motion is equal to the distance traversed in the same time by a body with uniform speed equal to half the final speed of the accelerated motion. Galileo takes AB (cf. Figure 5.4) to represent the time in which a body starting from rest at C traverses the space CD as its speed increases uniformly, and represents the greatest speed thus attained by the line EB, drawn perpendicularly to AB. He then draws the line AE and claims that all the lines drawn parallel to EB, from every single point on the line AB up to AE, represent the increasing degrees or "moments" of speed (an idea close to what we call the instantaneous velocity) of the body after the instant A. He next draws FG parallel to AB, with point F bisecting EB, and then points out that all the lines parallel to EB in the parallelogram AGFB and those in the triangle AEB represent, respectively, all the moments of speed in uniform motion and all the moments of speed in uniformly accelerated motion, during the same time AB. Galileo assumes (without argument) that all the speed momenta together (he speaks of their "aggregate," *aggregatum*) are proportional to the spaces traversed. Hence, he concludes, the equality of the aggregates of speed momenta entails the equality of the spaces traversed (Galilei [1638] 1954: 173–4).

Galileo derived from this theorem that distances traversed in free fall are to each other as the squares of the times of fall, or equivalently, that in equal intervals of time the distances increase as the odd numbers, 1, 3, 5, 7, . . . A crucial question remains: Do falling bodies actually undergo the acceleration Galileo described? Does the speed in free fall increase in proportion to time or rather in proportion to the distance already traversed, as was generally assumed until then? The young Galileo had entertained the latter view, as he explained in *Two New Sciences* (1638). There, he closed the discussion by appealing to an experiment with falling bodies on an inclined plane devised to demonstrate that his theorem describes actual free fall (Galilei [1638] 1954: 178–9).

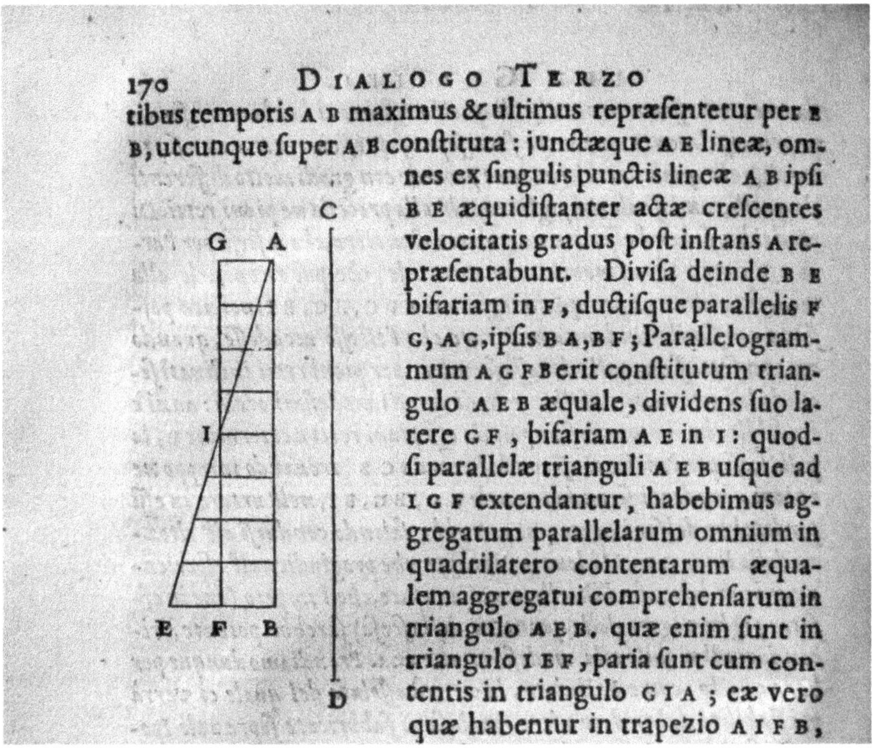

170 DIALOGO TERZO

tibus temporis A B maximus & ultimus repræſentetur per E B; utcunque ſuper A B conſtituta: junctæque A E lineæ, omnes ex ſingulis punctis lineæ A B ipſi B E æquidiſtanter actæ creſcentes velocitatis gradus poſt inſtans A repræſentabunt. Diviſa deinde B E bifariam in F, ductiſque parallelis F G, A G, ipſis B A, B F; Parallelogrammum A G F B erit conſtitutum triangulo A E B æquale, dividens ſuo latere G F, bifariam A E in I: quodſi parallelæ trianguli A E B uſque ad I G F extendantur, habebimus aggregatum parallelarum omnium in quadrilatero contentarum æqualem aggregatui comprehenſarum in triangulo A E B. quæ enim ſunt in triangulo I E F, paria ſunt cum contentis in triangulo G I A; eæ vero quæ habentur in trapezio A I F B,

FIGURE 5.4. The geometrical figures accompanying Galileo's proof of the mean-speed theorem, from his *Discorsi e dimostrazioni matematiche* (1638: 170–1). Bayerische Staatsbibliothek München: Rar. 4085.

Galileo concludes *Two New Sciences* by investigating the motion of projectiles and demonstrating in particular the parabolic nature of their trajectories. In doing so he stresses the assumption of a "new and subtle" argument, namely that the body answers to two motions (one horizontal and uniform, the other vertical and uniformly accelerated) and both "motions and velocities [. . .] combine without altering, disturbing, or hindering each other."

Galileo assumes a body is moving from *a* to *b* (cf. Figure 5.5) with uniform horizontal speed on an elevated horizontal plane that ends abruptly at *b*, at which point the body starts falling perpendicularly with uniformly accelerated motion, after which, says Galileo, its path will describe a semi-parabola *bifh*. The line *be*, prolongation of the plane *ab*, representing "the flow of time," is divided in equal intervals, *bc*, *cd*, *de*. From the points *c*, *d*, *e*, draw lines perpendicular to *be*. On the first, mark *ci* to represent the distance the body falls during the time *bc*; on the second mark *df* equal to four times *ci*; on the third, *eh* equal to nine times *ci*; and so on, their lengths always in proportion to the squares of the

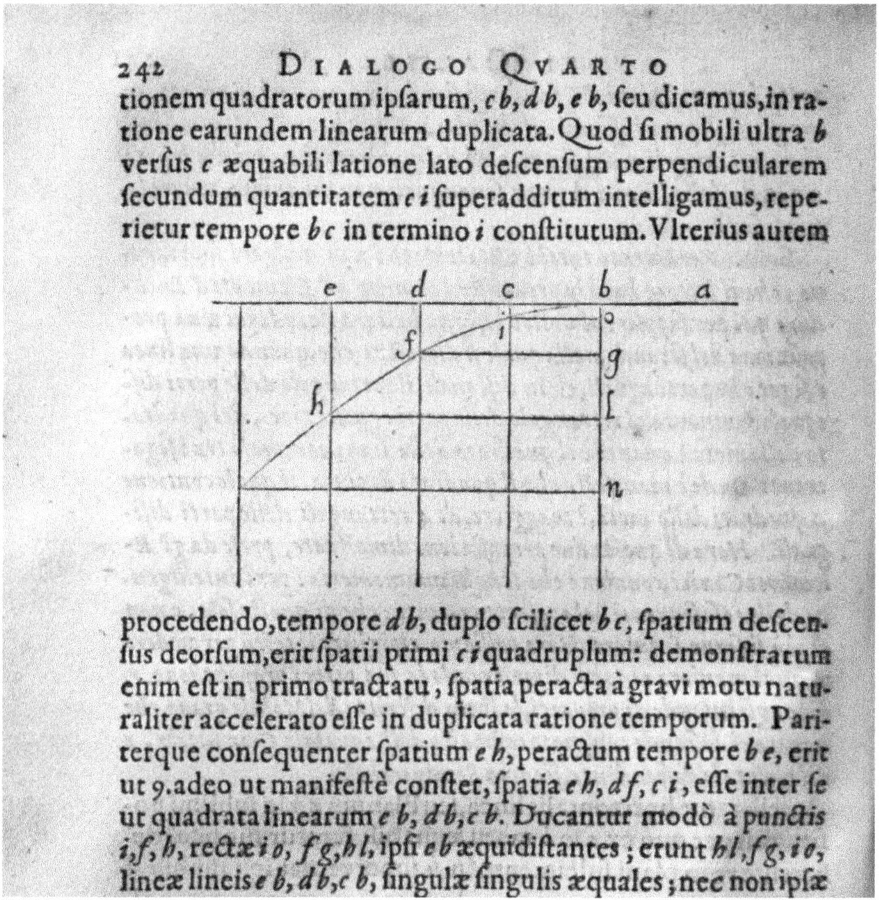
242 DIALOGO QVARTO

tionem quadratorum ipſarum, *cb*, *db*, *eb*, ſeu dicamus, in ratione earundem linearum duplicata. Quod ſi mobili ultra *b* verſus *c* æquabili latione lato deſcenſum perpendicularem ſecundum quantitatem *ci* ſuperadditum intelligamus, reperietur tempore *bc* in termino *i* conſtitutum. Vlterius autem

procedendo, tempore *db*, duplo ſcilicet *bc*, ſpatium deſcenſus deorſum, erit ſpatii primi *ci* quadruplum: demonſtratum enim eſt in primo tractatu, ſpatia peracta à gravi motu naturaliter accelerato eſſe in duplicata ratione temporum. Pariterque conſequenter ſpatium *eh*, peractum tempore *be*, erit ut 9. adeo ut manifeſtè conſtet, ſpatia *eh*, *df*, *ci*, eſſe inter ſe ut quadrata linearum *eb*, *db*, *cb*. Ducantur modò à punctis *i*, *f*, *h*, rectæ *io*, *fg*, *hl*, ipſi *eb* æquidiſtantes; erunt *hl*, *fg*, *io*, lineæ lineis *eb*, *db*, *cb*, ſingulæ ſingulis æquales; nec non ipſæ

FIGURE 5.5. The geometrical figure accompanying Galileo's proof of the parabolic shot from his *Discorsi e dimostrazioni matematiche* (1638: 242). Bayerische Staatsbibliothek München: Rar. 4085.

number of units of time marked on *be*. Since Galileo has proved that in free fall distances traversed vary as the squares of the time, he concludes by appealing to the known property of the parabola that the squares of its ordinates upon its axis are proportional to the distances on the axis from the ordinates to the vertex. Therefore the points *i*, *f*, *h*, . . . lie on one and the same parabola.

Galileo did not write *Two New Sciences* as a treatise on theoretical mechanics, however much it is now appreciated as the foundational work for this field. Since his primary concern was practical ballistics, Galileo may be regarded as Tartaglia's continuator. He discussed at length the issue of air resistance and whether his theory of projectile motion applies in practice. He contrived an ingenious experiment to compare the motions of pendula moving with different speeds, concluding that "the resistance of the air does not affect motions of high speed more than those of low speed." He claimed that, on the whole, his results may be regarded "without sensible error [. . .] as absolutely true" (Galileo [1638] 1954: 254). He closes his tract with propositions on the properties of parabolic trajectories and tabular results on them, including numerical values for parabolic shots.

Most of Galileo's results were recognized to be exact. His claim that the pendulum was isochronous, however, was soon contested and shown to be wrong. Galileo's emphasis on geometrical description versus causal explanation was also influential, and his work had a huge impact across Europe. In particular, it sparked sophisticated attempts to measure the distance traveled by a free-falling body in one second of time. In Paris in the 1640s, Marin Mersenne (1588–1648) contributed greatly towards publicizing Galileo's results. He and Giovanni Battista Riccioli, in Bologna, produced some of the most well-known and trusted experimental determinations of the constant of gravitational acceleration later known as g.

Christiaan Huygens (1629–95) became interested in Galileo's mechanics when he was a precocious teenager in the late 1640s, encouraged by his correspondence with Mersenne. The son of a wealthy upper-class Dutch family, Huygens was a mathematical polymath. In the 1650s, he worked on mechanics, optics, probability, observational astronomy (deciphering Saturn's odd appearances among other things), the measure of π, and also on horology. Huygens's interest in clocks was both theoretical and practical. He worked on the technical design of a pendulum clock accurate enough to be useful in calculating longitudes. In 1656, in collaboration with an Amsterdam clockmaker, he introduced a major innovation in the traditional verge escapement (which he patented in 1657) that strikingly improved the accuracy of pendulum clocks. Thanks to the innovation, the average daily deviation for clocks went down from 15 minutes to 10–15 seconds—an accomplishment that led to Huygens's reputation as the inventor of the pendulum clock. He published his invention in 1658 in a 15-page booklet entitled *Horologium* (Clock).

Yet, Huygens also had a theoretical interest in pendular motion. It was known that circular pendulums introduced substantial temporal deviations, when oscillations became larger than a few degrees. In answer to this inconvenience Huygens searched for the curve such that motion along it under the force of gravity will be isochronous—namely, such that the time to reach the lowest position will be independent of the height of the starting point. He assumed that pendulums oscillating along such a curve would be most accurate. Huygens also worked on conical (or circular) pendulums, and these investigations led him to a major breakthrough: he found the mathematical expression for "centrifugal" force (as it was then called).

Huygens made use of infinitesimals, then a mathematical innovation. Infinitesimals, which had been making occasional appearances since the turn of the century, had by the 1650s been established through the works of Cavalieri, Torricelli, Pascal, Roberval, Wallis, and others (cf. Chapter 3). Huygens took instantaneous speed to be the *uniform* speed a body has during an infinitesimal interval of time. A few years later, Newton (independently) grounded his fluxional calculus in the same idea. Huygens's insight was to examine pendular motion in its infinitesimal or "local" elements, where crucial simplifications would apply. Huygens's ability to analyze pendular motion during an infinitesimal interval of time, combined with his intimate knowledge of the geometry of the circle and conic sections, allowed him to recognize the isochronous nature of the cycloidal pendulum, namely a pendulum whose bob describes a cycloidal rather than a circular arc.

The cycloid was a fashionable subject of discussion among mathematicians in the 1640s and 1650s (as described in the volume Introduction). This curve is generated as the path described by a point on the circumference of a circle rolling along a straight line—in Figure 5.7, as the circle EMK rolls without translation along the straight line AN, the point E (originally at A) describes the cycloid, AEFN. Once Huygens knew the isochronous nature of the cycloid, he had to devise a way to make a pendulum bob move along a cycloidal arc. In 1657, he had used metallic plates (called "cheeks") to constrain the motion of the pendulum's suspension chord in order to prevent large oscillations (cf. Figure 5.6). In 1659 he addressed the question, for which curve ABC (cf. Figure 5.7) will the suspension chord, EKBC, when wound around the plate CB, cause the bob E to describe the cycloid AFN?

Given a convex line such as ABC (cf. Figure 5.7), Huygens imagined a string stretched around it with one end remaining fixed at C (here represented partially unwound). If the other end is pulled away so that the string remains taut, then this free end describes a curve, AEF, that Huygens called the curve described "*ex evolutione*," literally "by unfolding." He called the curve ABC the *evoluta*, "the unfolded one," now called by the somewhat deceptive cognate *evolute*, while the curve AEF is now called the *involute* (Huygens did not attach a name to it, he always called it the curve "described *ex evolutione*"). By construction, as E moves along the curve, the line EB is always tangent to the evolute at B, and normal to the curve AEF. Assuming AEF to be a cycloid, Huygens looked for its evolute.

Again, infinitesimal thinking, allied with Huygens's knowledge of the cycloid, was instrumental in his discovery. Starting from AEFN (cf. Figure 5.7), he took two infinitesimally close normals to the curve at E (only one is represented here, EK) and determined the point where they met. Huygens reasoned that this point (B) was the point of intersection of two infinitely close tangents to the curve to be determined (ABC), and would therefore be on the curve which he sought. He then demonstrated that EB was twice EK, and drawing on his intimate knowledge of the cycloid, he proved that B belongs to a cycloid generated by a circle AHD equal to the generating circle of the cycloidal path AEFN.

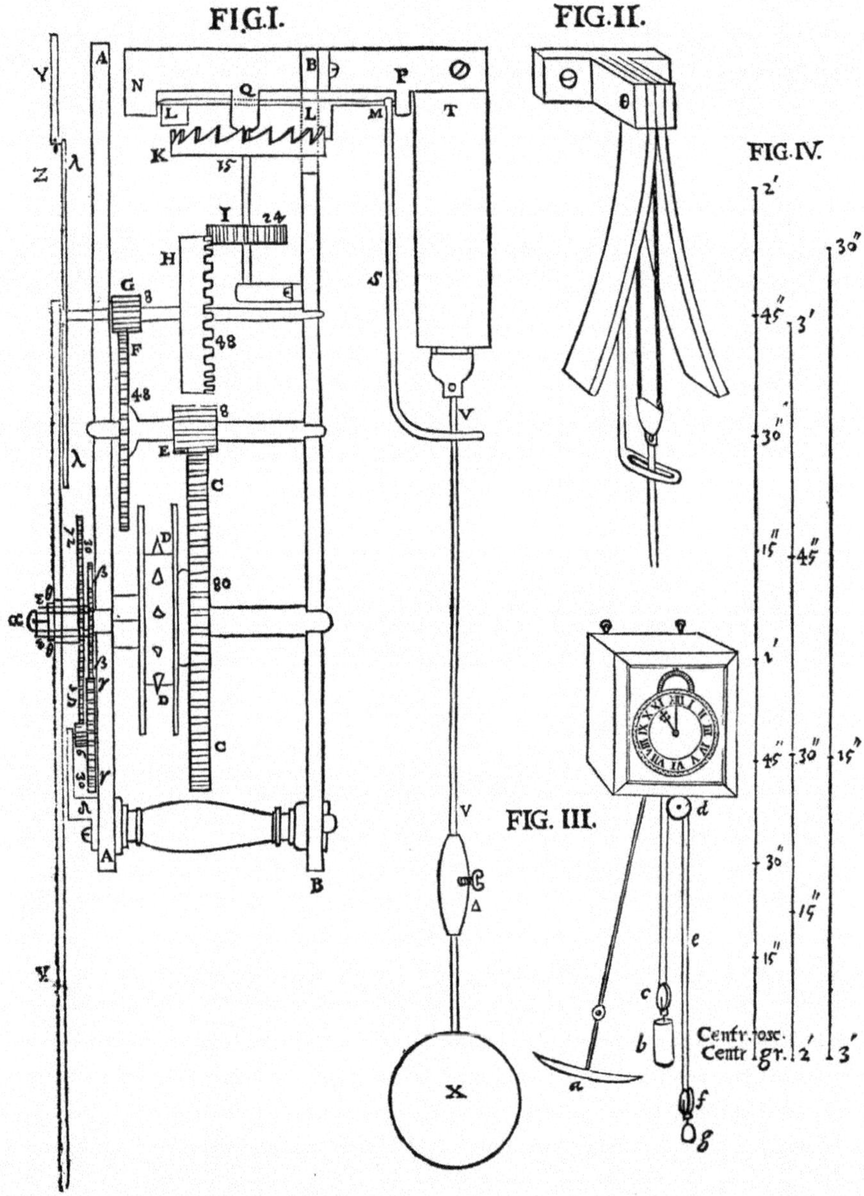

FIGURE 5.6. Huygens's pendulum clock (from *Horologium oscillatorium* (1673: 4). The cheeks are reproduced under the caption "FIG. II". They are represented sidewise in "FIG.I", marked T. Photo by Photo 12 / Universal Images Group / Getty Images.

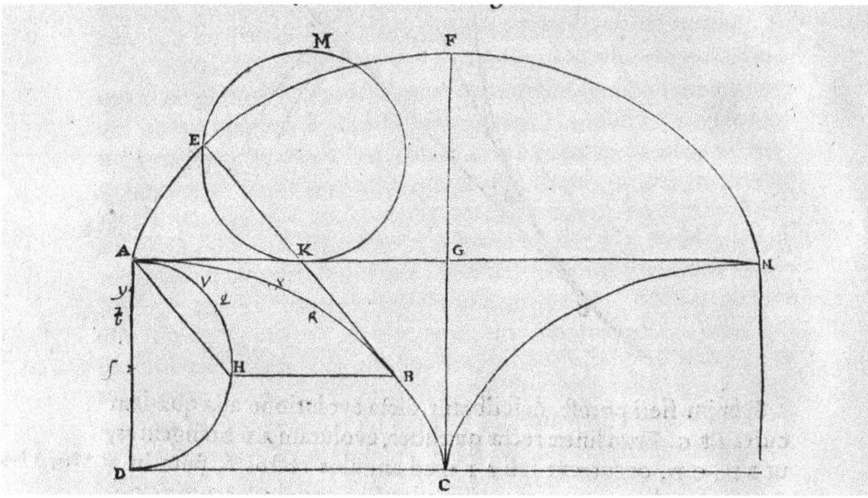

FIGURE 5.7. From *Horologium oscillatorium* (1673: 66). Gottfried Wilhelm Leibniz Bibliothek.

Since the length of the pendulum, CF, was twice the diameter of the generating circle, AD, Huygens could have the cheeks for a pendulum precisely manufactured by giving them the shape of a cycloid generated by a circle whose diameter was half the pendulum length. He assumed a length of 9 and a half inches for a 1 second pendulum, and built a clock with cheeks made from a cycloid generated by a circle whose diameter was half that length. Huygens hoped the new pendulum clock, grounded on the laws of motion and equipped with geometrically designed cheeks, would be a breakthrough in horology, in particular for seafaring purposes. In such a clock, irregular oscillations were no longer a source of error—regardless of the amplitude of the swing, the pendulum beats the same time.

On the mathematical side, Huygens exploited the "evolution" of curves to attack the problem of rectifying (finding the length of) curves. He found the evolute of a parabola and then looked for, and found, a general solution for determining the evolute of geometrical curves that was nearly equivalent to the calculus-language modern formula for the radius of curvature. He also compiled a list with many (then) common curves and their evolutes, including the conics and higher order parabolas.

All this wealth of theoretical results and technological innovations, and more, was eventually published in 1673 in one of the greatest mathematical and mechanical treatises of the seventeenth century, the *Horologium oscillatorium* (The Pendulum Clock or Geometrical Demonstrations Concerning the Motion of Pendula as Applied to Clocks; Huygens 1986). The results Huygens published here constitute one of the most difficult and fascinating set of problems solved with pre-calculus techniques. Huygens's intellectual success in solving the isochrony problem and implementing this solution by means of cycloidal cheeks was not accompanied by practical success. These new clocks proved much too expensive for land use, as the gains in accuracy were minimal compared with circular pendulums with very small oscillations. On the other hand, sea trials showed that no pendulum clock could keep time accurately due to the uneven motions of seventeenth-century vessels on high seas.

The uses of mathematics in early modern mechanics paved the way for one of the greatest achievements in seventeenth-century natural philosophy, Newton's mechanics. It is remarkable that here, as was the case in cartography, mathematics had to solve problems that required ideas and methods which had not yet been properly defined (integrals, limits, infinitesimals). Problems of great difficulty then appeared, which very much enriched mathematics. Yet, it happened more than once (e.g., with Gregory, and Huygens, as we saw) that once the problems were solved, the solutions held no interest from a practical point of view, although they were endowed with tremendous theoretical import.

"POLITICAL ARITHMETIC" AND GAMES OF CHANCE

In the second half of the seventeenth century, along with the growing prestige and enhanced social status of natural philosophy and the mathematical sciences, the idea emerged of applying empirical and "philosophical" principles (which were not bound by tradition) to the governance of the state. William Petty (1623—87) stands as a case in point. A physician by training, member of the Paris-based Mersenne circle and follower of Thomas Hobbes in the 1640s, he turned in the 1650s to public service, becoming chief surveyor of Ireland for the army of Oliver Cromwell—which made him enormously rich. A founding Fellow of the Royal Society of London, he contributed a few natural-philosophical papers, but from the 1660s on his main, highly innovative preoccupation was how to apply mathematical arguments to policy matters—what he called "political arithmetic," which is the actual title of one of his books (first published in 1690, Petty [1899] 1986). He wrote about population growth, taxation, immigration, trade policies, and the evaluation of the nation's wealth, but also about the real sources of wealth and how to promote them. His rationalistic perspective and mathematical acumen put an economic focus on problems that had never been thus analyzed. He estimated a man's monetary value by the revenue he produced, and contrasted this with the cost of slaves. He also demonstrated the benefits of religious toleration with economic arguments and "political arithmetic" and showed that it is socially more profitable to fine a robber, or enslave him (were he an indigent), than to hang or mutilate him. Most of what he wrote regarding taxation and topics in (what is now called) political economy is considered original, innovative and well founded by historians of economic thought, and he is in fact considered one of the founders of the field. Although he was close to members of Restoration England's political elite and well regarded by Charles II, his advice and proposals went largely unheeded, to his great disappointment. Petty's abstract, rational, mathematical approach must have clashed with the constraints and uncertainties of what was considered politically reasonable (Olson 1990: vol. 2, 72–86). Forty years after Petty's death, his mathematical approach to governance and in particular his proposals for the prosperity of Ireland were bitterly ridiculed in Jonathan Swift's *Gulliver Travels* (Book III) and *A Modest Proposal for . . . Poor People in Ireland.*

The mathematics Petty applied in his quantitative studies of demography and political economy was hardly innovative as he essentially relied on averaging and extrapolation. Eventually the quantification of social and policy phenomena would benefit from contemporaneous studies of chance events, which did involve highly innovative mathematics. Mathematical probability took shape in the seventeenth century in intimate connection with the analysis of problems arising in games of chance, insurance, and annuities. The first important ideas can be found in the 1654 correspondence between Blaise Pascal (1623–62) and Pierre de Fermat (1607–65) about the fairest way to divide the stakes between two players, if their game is interrupted. The letters (not published until 1679) kept the focus on games and contain no systematic, abstract treatment of the concepts and techniques involved.

Huygens, who knew the correspondence between Pascal and Fermat, published the first treatise presenting a theory of chance events, *De ratiociniis in ludo aleae* (On Reasoning in Games of Chance, 1657). Nonetheless, the first formalizations of probabilistic thinking were indebted to the legal tradition of aleatory contracts, or contracts wherein agreements involved an element of chance. In particular, the legal notion of an equitable business contract, one in which all partners "fairly share the risks" and have a fair "expectation" of benefits, was highly influential. It was in terms of *expectations* rather than probabilities that Huygens wrote his tract, which facilitated its application to an actuarial problem with political implications (Daston 1988; Hacking [1975] 2006).

Johann de Witt (1625–72) belonged to the Leyden circle of young mathematicians tutored by Frans van Schooten (1615–60), who was also a mentor for Huygens and translator of Descartes's *La Géométrie* into Latin. The son of a prominent family in the Netherlands, de Witt was Grand Pensionary from 1653 until his assassination in 1672, with responsibilities equivalent to those of a prime minister. Out of his concern for improving public finances came in 1671 his *Waerdye van Lyf-Renten* (Treatise on Life Annuities), a very early application of probabilistic notions to insurance problems.

Annuities were a standard tool for public institutions to raise money. A life annuity on someone (usually a healthy child) entailed paying a lump sum in exchange for the beneficiary receiving an annual income for life. Drawing inspiration from Huygens's book, de Witt assumed that when a person has a chance of winning A and an equal chance of winning B, he has an *expectation* of winning (A+B)/2. He further assumed uniform rates of mortality for the age intervals 4–53, 54–63, 64–73, and 74–80, namely that the chances of dying are equal in all the years of each interval. He assumed as well that the chances of dying in these intervals are in the ratios 1:3/2:2:3, namely a 74-year-old is three times as likely to die as a 4-year-old, and so on. On these assumptions, de Witt demonstrated that the Dutch republic could sell annuities at 16 years' purchase (1 florin of annuity would cost 16 florins) rather than at 14 years' purchase, as was customary, and they would still be profitable to buyers. His argument was that investors could mathematically expect the annuity beneficiary to receive by the end of his or her life at least as much as he or she would have gotten if the money were lent at 4 percent, the usual lending rate.

De Witt's analysis shared the fate of Petty's proposals. It was not put into practice, nor was probability theory put to use in public policy decisions for decades to come. De Witt died before he had time to implement his ideas, but implementation would have posed major technical problems, not to mention strong social resistance. The main technical problem was the lack of reliable tables of mortality rates, a problem that took many years to overcome. De Witt's and Petty's ideas nicely exemplify a trend in those seventeenth-century mathematical results that were developed as tools for practical ends, be they economic or political, cartographical, or mechanical. Early modern mathematics proved crucial in advancing astronomy, mechanics, and cartography, in shaping the fine arts and improving optical theory and instruments, and in conceptualizing probabilities. However, deeply original and powerful results often proved impractical because they were too sophisticated for what was actually needed, or simply because they were impossible to apply given the social context or the available technological skills. The fact remains that more often than not the efforts to use mathematics to understand and describe nature repaid enormous benefits for pure mathematics itself.

EARLY MODERN MATHEMATIZATIONS OF NATURE

Since Antiquity and through the Middle Ages, mathematics was intensively applied in astronomy and optics, and to a lesser extent in mechanics and cartography. Leaving astronomy aside (which dealt with heavenly phenomena, different in kind from terrestrial ones within the predominant Aristotelian framework), Euclid's optics and Archimedes's statics offered paradigms of mathematization of physical knowledge. They were examples of Aristotelian "mixed" mathematical sciences. Mathematization in this context meant assimilating physical objects (visible points, light or visual rays, scales arms, weights) to geometrical objects (geometrical points and bodies, straight lines, etc.). This involved expressing indubitable physical postulates in a strictly geometrical idiom. For instance, in optics it was postulated that visual rays emanate in straight lines from the eye, and that magnitudes observed under greater angles appear greater; in statics it was postulated that equal weights at equal distances on the arms of a lever are in equilibrium, and so on. The mixed mathematical sciences of Euclid and Archimedes were in style, method, and organization closely similar to Euclid's *Elements*. Besides the axioms and postulates of the *Elements*, mixed sciences took for granted obvious physical (namely, with empirical content) axioms and principles (geometrically expressed). These were used to ground Euclidean demonstrations about weights and the appearances of material objects.

In the early mathematizations of Euclid's optics and Archimedes's statics, the theorems express something necessarily true about the physical world, since they are necessary conclusions grounded on physical truths. Checking empirically, for instance, that weights inversely proportional to the lengths of a scale's arms are in equilibrium, can only have a pedagogical (or similar) purpose, not a demonstrative one. The empirical truth of geometrically demonstrated results in such mixed mathematical sciences is guaranteed by strict geometrical reasoning grounded on empirical truths. Demonstratively speaking, empirical evidence is redundant.

The classical model of the mixed mathematical sciences embodied in the works of Euclid and Archimedes regained centrality in the sixteenth and seventeenth centuries. Innovative, influential works by Stevin, Pascal (on hydrostatics), and Huygens were written very much in this style and with the logical structure of such mixed mathematical sciences. In Stevin's statics and hydrostatics (printed in Dutch in 1586 and 1608, in Latin in 1608, in French in 1634), the objects studied (weights, balances, and so on) are defined in purely geometrical terms. He then formulates a few postulates deemed to be obviously true—equal weights at equal distances are in equilibrium, all weights act downward on parallel lines, water is of uniform gravity throughout, the upper surface of a body of water at rest is planar and parallel to the horizon. From the postulates, theorems follow along the lines found in Euclid's *Elements*. Stevin's works have been considered the natural successors of those of Archimedes and their methodology differs in no way from the latter's.

Huygens's treatise on the laws of impact (*De motu corporum ex percussione*, *c*. 1656, printed 1703) is a highly innovative, conceptually sophisticated treatise dealing with "perfectly hard" (now called perfectly elastic) bodies. In it, Huygens highlighted three "hypotheses"—the principle of inertia, the relativity of motion, and the principle that equal bodies moving with equal and opposite velocities, after collision rebound with the same velocity they had but opposite in direction.

At some point, an additional "axiom" ("very certain [. . .] in mechanics") was brought into play, namely that for bodies moving by gravity alone, their center of gravity cannot ascend. Mathematics did the rest, enabling Huygens to demonstrate in 13 propositions the results concerning all the instances of the impact between two bodies with a style "similar to that of Archimedes" (Dugas [1955] 1988: 180).

The ideal of classical mixed mathematical sciences was also reflected in the popularity of the term "physico-mathematics" during the middle decades of the seventeenth century, when it appeared on the title pages of substantial works by Marin Mersenne, the Jesuit mathematician Paul Guldin (1577–1643), and others. Physico-mathematics, which gained currency in both vernacular and Latin works and particularly (but not only) among Jesuit mathematicians, was not just a new name for mixed mathematical sciences. It applied to new knowledge being produced in magnetism, pneumatics, mechanics, etc., in a way that promoted the idea that mathematics yields causal knowledge about natural phenomena—a viewpoint that did not fit into Aristotelian or Scholastic ideas of natural philosophy (Dear 1995). Physico-mathematics elevated the philosophical status of mathematics, while embodying a process that has been called "disciplinary imperialism," namely the process by which mathematics takes over matters that had traditionally been part of physics.

In Galileo's kinematics as well as in Huygens's analysis of pendular motion, the classical ideal of applying mathematics to physics by rigorously following the axiomatico-deductive method began to fall apart. For one thing, it is difficult, albeit not impossible, to introduce deeply innovative ideas and results starting from plain, obvious, non-contentious truths. For another, innovative authors were stepping out of the well-trodden paths of Euclidean geometry to embrace new, and contentious mathematical objects—indivisibles, infinite series, new curves, and so on. We find philosophical articulation of the crisis of the mixed mathematical sciences in the *Mathematical Lectures* (printed 1683, delivered 1664–6) of Isaac Barrow (1630–77), the Lucasian Professor of Geometry at Cambridge who was lending books to the young Newton when he was an undergraduate there. Barrow advocated the mathematization of nature as the best approach to natural philosophy in general, not just to a few parts of it. Then he introduced new viewpoints on the nature of axioms, definitions, and first principles. He dropped the requirement of universality and self-evidence for mathematical principles opting instead for the weaker demand that they merely be "free from contradiction." Mathematical results built on such principles, he added, may or may not apply to our natural world. Yet, they can be mathematically true even if they "do not apply to this world." Barrow's remarkably innovative ideas were applied to optics in his *Geometrical Lectures* (1670), where he used sophisticated geometry to determine the location and shape of images produced by lenses and mirrors, claiming that his theoretical results could be tested by means of observations.

Early modern mathematics substantially expanded its natural philosophical dominion while gaining philosophical authority and institutional ascendancy. It was crucial that new mathematical tools (infinitesimals, new curves, power series, etc.) enabled natural philosophers to pose new problems and to imagine new answers for old ones. These new mathematical tools and methods contributed to weaken the haunting spell of the axiomatico-deductive order. It was an order characteristic of the much admired organization of Euclid's *Elements*, but it was also the informal pattern characteristic in general of the Aristotelian theory of knowledge. Sound, necessary, scientific knowledge required principles of unquestionable certainty, followed by sound rigorous reasoning. It was a pattern that pervaded not only "physico-mathematics" and natural philosophy but also the great contemporary philosophical systems—including those of Descartes and of Hobbes. Its final crisis would only come in the wake of Newton's *Principia mathematica* (1687) and the advent of the infinitesimal calculus (cf. Chapter 3 in this volume and in Volume 4).

CHAPTER SIX

Mathematics and Technological Change

The Silent Rise of Practical Mathematics

THOMAS MOREL

The role of technology changed considerably in the early modern period. Conflicts and epidemics had plagued Europe in the late Medieval Age, and the somewhat calmer recovery period correlated with an accelerated rate of technological innovation. This development, marked in Western and Central Europe, exerted a crucial influence on a global scale due to overseas exploration and subsequent colonization. As a reference work on the history of technology puts it, "the technical level that was reached and mastered put the European States in the position to define the course of World history" in the coming centuries, with far-reaching consequences (Hägermann et al. 2003: 210).

The exact role played by mathematical sciences in this process, difficult to assert as it may be, will be the subject of this chapter. A few preliminary remarks should be made, however, in order to qualify any further claim about the relationship between science and technology in the early modern period. First of all, most of the technical fields were free from immediate mathematization or from any scientific formalization of knowledge. In other words, most early modern technicians and artisans never considered using mathematics, be it numerical tables or technical drawing, simply because it could not be expected to be useful in any way.

This was true for many prevalent activities of the time such as agriculture, mining, textile manufacturing or dyeing and tanning, but also for innovations such as the printing press, whose development did not involve formal exact knowledge. Even technical achievements such as the medieval cathedrals were based on craft knowledge that "was almost completely non-mathematical. It permitted a great many variations to be played on a basic set of geometrical forms" (Knobloch 1994: 251). It would be hard to overstate how much this approach differs from our modern world, where the relationship between mathematics and technology has been deeply reshaped during the Industrial Revolution. This observation corroborates, in the technical world, conclusions drawn in the previous chapter: in the early modern period, mathematizing nature was still an ambitious project, not yet a general reality (cf. Introduction and Chapter 5).

An important change that occurred at the beginning of the fifteenth century was the rise of a writing culture among craftsmen. Beginning with a growing number of practical treatises by artisans in the first half of the century, this process took on another dimension when the printing press with movable types was developed and disseminated. It is thus crucial to emphasize the role of skilled artisans when discussing mathematics and technological change. Building on the intuition of Edgard Zilsel, the modern historiography emphasizes the key role played at the time by artists-engineers, also known as mathematical practitioners (Cormack, Walton, and Schuster 2017).

That being said, the early modern period experienced tremendous transformations in some specific domains. While mathematics is often depicted as an abstract science, it was at the time gaining a wider audience in civil life. Arithmetic was used not only by wine or beer gaugers but also by merchants and bankers. The efficiency of geometry was most visible in the works of surveyors and in the portolan charts used by navigators (cf. Chapter 5). While most of these practitioners had presumably never studied theoretical geometry or read the second and sixth books of Euclid's *Elements*, they increasingly "felt themselves to be part of a thriving enterprise that had proved itself in a number of areas and was poised to extend into others" (Bennett 1991: 181). And yet, other historians come to the seemingly opposite conclusion, i.e. of an "absence of a science-based technology in the seventeenth century," simply because "the gap between a solution that worked in theory and a solution that worked in practice proved to be far greater than could be overcome at the time" (Cohen 1994: 193, 342).

Why such a discrepancy in appreciating the use of mathematics by early modern technicians? It mainly comes down to the versatility of what "mathematics" meant (cf. Introduction), and to its uneven applicability to various technological fields. This chapter uses the category of "practical mathematics" to denote knowledge mainly produced by technicians, engineers, artists or merchants for their own practical needs. In other words, it forms a disparate set of useful formulas, methods, or techniques used in civil life. While these persons cared about accuracy and steadiness, little attention was given to abstraction or demonstration. This use of mathematics thus differs markedly from both our modern conception of science-based technology and from the contemporary *quadrivium*. There was an undeniable rise of numbered information in the early modern period, with the most diverse audiences absorbing the popular saying from the *Wisdom of Solomon*, according to which God "hast ordered all things in measure and number and weight." However, this general movement was mostly uncoordinated, and its influence on technological change cannot be evenly accounted for. It could indeed take very different forms.

Altogether, uses of mathematics in engineering were met with mixed success. In some areas such as surveying, mathematics was a driving force and could already be used as a tool for innovation in order to discover or implement new methods. In some fields, available mathematical methods could hardly influence technological change despite numerous efforts, as our first case study on waterwheels will illustrate. In most cases however, it is difficult to find a proper scale to evaluate the success of repeated attempts made by both practitioners and scholars—albeit in substantially different ways—to produce formulas, plans, tables, diagrams, and other mathematical tools aiming at improving practices. Mathematics could namely play a significant role in technological change without driving it, as will be shown in various examples.

Apparent contradictions about the influence of mathematics ultimately boils down to diverging views of what "mathematization" could mean in the early modern period. Nowadays, we mostly think of it as engineering, that is, a reliable modelization of a problem, based on which calculations and theoretical investigations can be performed to produce reliable forecasts or improve existing practices. In the early modern period, mathematization would usually be more sketchy and gradual, without an obvious overarching roadmap. Tables built on experience or misnamed "rules of thumb" about proportions can be seen as very effective first steps towards mathematization, provided that they explicitly use numbers and codified procedures.

More ambitiously, one can think of sets of formulas and rules relative to a profession that were empirically discovered and gradually improved, adapted, and generalized. During the Renaissance, many examples of mathematization would fall in that intermediary category. Using three-dimensional models made of wood or plaster was useful for visualization, but the problem of scalability did not allow for direct transpositions from these into reality. Full-scale buildings would sometimes crumble under their own weight, even when corresponding models had seemed perfectly solid. Well aware of this caveat, engineers would still rely on models for lack of better tools such as our modern stress analysis, adjusting with their experience and rules of proportion.

Uses of mathematics gradually changed from the fifteenth to the seventeenth century, as society at large underwent deep historical transformations. The early modern period was characterized by the rise of states and administrations, evolutions that would have a deep impact on the use of mathematics in warfare. On an individual level, the development of mathematical education made low-key mathematization possible (cf. Chapter 1). While technological prowesses are often related to major figures such as Leonardo da Vinci (1452–1519), Albrecht Dürer (1471–1528), or Simon Stevin (*c.* 1520–1620), a growing number of lesser known technicians came to master enough mathematics to have a significant influence on practices. In many cases, mathematics was brought forward neither by scholars nor in universities, but increasingly by practitioners, artisans, and engineers. The engineer is indeed a key figure of the early modern period, heavily relying on mathematics to "ease the discovery of new directions in which to look for, in order to find the most appropriate solution" (Vérin 1993: 16).

The following sections present a nuanced view of the impact mathematics had on technical activities. Some classic and well-known examples are given, together with more specific or esoteric case studies, in order to underline the diversity of actors, methods and possible outcomes. I deliberately begin with the unsuccessful attempts to improve the design of waterwheels using mathematical theories. I then turn to the various uses of surveying. From Dutch polders to German mines, and from the draining of the English Fens to mapmaking, surveying provided some of the most tangible successes for early geometry. A growing use of tables, data collection, and formulas was a much less visible evolution, but its impact ultimately proved even more transformative. Finally, a selection of major architectural enterprises will underline the growing scale of engineering projects developed in the burgeoning nation states of Western Europe.

IMPROVING THE WATERWHEEL: A FAILURE OF MATHEMATICS?

Technological change is to a large extent dependent on available energy sources. From the end of the medieval period, human and animal power was increasingly replaced by waterpower. Elaborate power systems were introduced, with dams, reservoirs, gates, and most importantly waterwheels (Reynolds 1983). Although their use can be tracked back to antiquity, both in European and Chinese sources, they only became ubiquitous during the late Middle Ages and the early modern period.

Watermills were erected everywhere, in cities and in the countryside, in lowlands and in the mountains. Most rivers were, wherever possible, cluttered by dozens of wheels, so that throughout Europe there were tens of thousands of mills. The universal concept was adapted to diverse geographical and economical settings, becoming a ubiquitous source of energy: wheels could power sawing mills, paper mills or flour mills, to name but a few. During the sixteenth century, large waterwheels were introduced by German engineers in the ore mines of the Ore Mountains and by Spanish *conquistadores* in the silver mines of Potosi (now Bolivia). Waterpower made deep-level mining possible, but was also used to crush the ore and power the furnaces. It worked on small-scale projects but could be adapted to any possible need, reducing labor and increasing reliability.

A waterwheel can be designed in several ways. It can be set horizontally or vertically, the latter being generally more efficient. Among vertical waterwheels, undershot wheels rely on the impulse of water coming from below to move the blades, while in overshot wheels it is mainly the weight of water that matters (many combinations and variations do exist, such as breast wheels or boat mills, and have been studied in detail; Reynolds 1983). Although it costs more, an overshot wheel can be roughly twice as efficient as an undershot wheel. In spite of the diversity of local circumstances, most engineers and millwrights had an empirical knowledge of this fact. Scholars, however, struggled to produce a quantitative analysis coherent with observed nature. The difficulty of hydrodynamics combined with the necessity of using calculus prevented any comprehensive mathematical analysis before the late eighteenth century (cf. Chapters 3 and 6 in Volume 4). Even basic computations to estimate the efficiency of waterwheels were wanting. The choice between undershot and overshot wheels was just one of many questions where theory could have brought (and later did bring) more precise answers than experience alone. What was the optimal number of blades? Which blade shape ensured the best use of water? What was the optimal speed of incoming water, or the most efficient flow rate?

However, no mathematical theory was able to answer these questions in the early modern period, and the few attempts produced conclusions that contradicted observations so blatantly that they were soon discarded. It might seem unfair to present the waterwheel as a "failure" of mathematics to drive technological change, when needed theoretical tools were simply not available. From an early modern perspective, however, waterwheels were a widespread power source whose improvement was badly needed. Many rivers were simply crowded with mills. With the primary resource—water—being limited, there was a clear incentive to improve the efficiency by all means. Given the technical and economical importance of waterwheels, they were a popular writing topic. Georgius Agricola (1494–1555) represented them among his mining machines in the *De Re Metallica* (On the Nature of Metals), and Jacques Besson (*c.* 1530–72) drew them along several engineering tools in his *Théâtre des instruments mathématiques* (Theater of Mathematical Instruments).

And yet, very little is known about the design and construction of waterwheels, except that generations of artisans slowly improved their average efficiency. Millwrights did not write their practices down, making any trace of mathematization hard to perceive. We know that the beautiful waterwheels depicted in the *De Re Metallica* were built by mining masters such as Martin Planer (1510–82), but Planer never published on the topic. Such wheels were built on a regular basis, improved and even combined. One can thus assume that practical knowledge was developed and transmitted, but testimonies about the use of mathematics are absent. There are hardly any sketches by early modern practitioners left, and these usually do not include scales nor precise proportions, so that they could in all likelihood not be used as technical instructions. In the mid-eighteenth century, John Desaguliers (1683–1744) had an ambivalent view of millwrights' mathematical knowledge. On the one hand, he respected the observation skills of gifted engineers, defining it as "a strong natural Genius for Mechanicks, that without any previous Knowledge of Mathematicks and Philosophy, have performed great Works," labelling this empirical approach "natural Mathematicks of their own," but then added:

> Almost all the Plumbers and Mill-Wrights now set up for Engineers; tho' I hardly believe that there are two of them who know how to measure the Quantity of Water requir'd to turn an Undershot, or an Overshot, or a Breast-Mill. They only judge of a Stream by the Eye, and he that has most practice is likely to succeed the best. (Desaguliers 1744: 414)

The scarcity of sources is all the more problematic given that some of the remaining documents support the idea of a culture of practical geometry among millwrights, at least in some of the most literate parts of Europe. In the Low Countries, these engineers were called *molenmeester* and regularly combined this office with the duties of surveyor or gauger. The various uses of mills and wheels led to significant improvements. In the draining and maintenance of polders, associations of waterwheels and windmills placed at different heights were developed in the early seventeenth century, in order to regulate the varying flow of water (cf. Figure 6.1). In a technical drawing made by Jan Adriaensz Leeghwater (1575–1650) representing the Beemster polder, one clearly sees that scales and precise measurements were made to ensure maximum efficiency. This use of mathematical tools, however, did not rely on any theory of hydraulics but on the experiences gathered by Dutch engineers.

Among the earliest notes about the design of waterwheels, one finds writings on hydraulics by Leonardo. These working notes can be contradictory or obscure, sometimes illustrated with sketches. In one instance, Leonardo studies "the true power of a quantity of water which falls above the wheel of a mill through varying angles, but from the same height to the same depth and on the same type and size of wheel," concluding that the impact was proportional to the height (Codex Atlanticus, 209r, as quoted by Reynolds 1983: 197). He gave

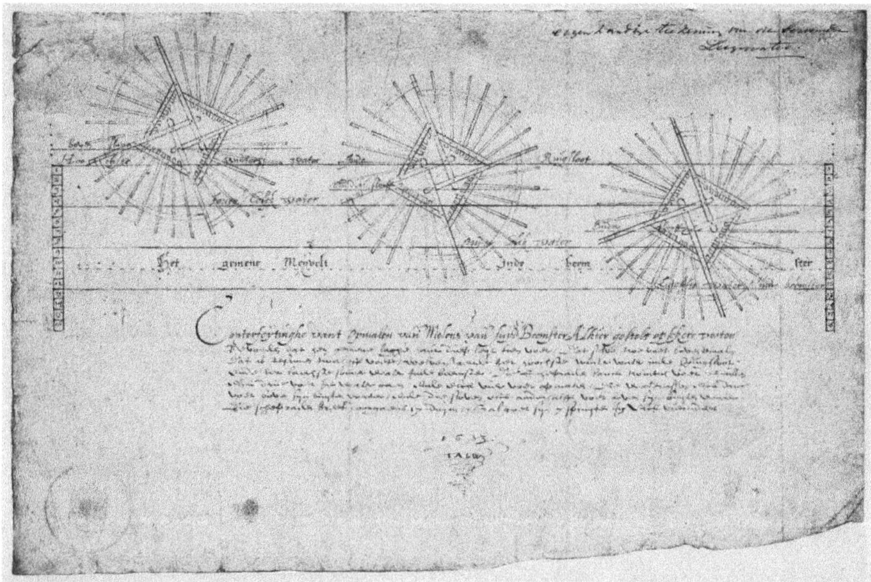

FIGURE 6.1. Jan Adriaensz. Leeghwater, succession of three waterwheels (powered by windmills) to draw the water from the lowest to the highest level (copy made in 1633). Noord-Hollands Archief: IALW A(492.631.010)029.

several numerical examples, but they did not draw on experiments and were only meant to illustrate the rules he had proposed.

The problem Leonardo encountered was not merely that his rule was false, but rather that the study of machines relied on numerous physical concepts that had to be disentangled before any mathematical analysis could take place. A less prominent member of this group of Italian artist-engineers of the fifteenth century is Francesco di Giorgio Martini (1439–1501), who wrote two treatises on architecture and machines. His studies on mills present the reader with minute descriptions of many variations, even giving formulas to calculate the "capacity" of mills relative to their axle. He did not design or compute new models based on theory, but relied on a collection of existing references. His *Trattato I* lists fifty-eight mills in total with precise dimensions for the diameter, the gears, axles, number, length, and thickness of the blades. However, the author underlines several times that every situation requires a specific adaptation and that practical experience is absolutely necessary to cope, for instance, with local construction materials. Leonardo aside, we thus see that artists-engineers possessed a sense of numbered information as well as catalogues of machines, but were not necessary looking for general and theoretical principles about the running of waterwheels.

During the seventeenth century, Italian scholars wrote extensively on mechanics, building on Galileo Galilei's (1564–1642) work. Given the theoretical and practical importance of hydraulics problems, they quickly became an important component in the *Académie des sciences* in Paris. Both Christiaan Huygens (1629–95) and Edme Mariotte (*c.* 1620–84) performed experiments. Their underlying goal was less the improvement of wheels than the production of formulas allowing for a comparison between human and waterpower. But even an ambitious treatise such as Mariotte's *Traité du mouvement des eaux* (The Motion of Water, 1686) could hardly have any practical impact, as it was restricted to the abstract case of a static waterwheel.

The key figure in the analysis of the waterwheel was Antoine Parent (1666–1726), who worked at the *Académie des sciences* and developed Mariotte's analysis in a dynamic setting. In a milestone 1704 article, he used calculus to demonstrate the ideal conditions under which a wheel would have the maximum effect. Building on this work, Bernard Forest de Bélidor (1698–1761) proved in his *Architecture hydraulique* (Hydraulic Architecture) that an undershot wheel should be six times more efficient than a similar overshot wheel. This and other findings concerning the number of blades or the appropriate speed of water were useless in practice, insofar as everyday experience easily proved them to be completely false. The complexity of mathematical modelization and the necessity of ignoring certain variables meant that purely theoretical analysis resulted in recommendations that were ineffectual or not practicable in the real world. Despite countless attempts to use mathematics to improve the design of the wheel and the shape or disposition of the blades, theory was not able to do better than the trial-and-error methods used by skilled practitioners.

In the long run, only comparative experiments with large-scale models proved able to produce theoretical predictions that matched with reality, as John Smeaton (1724–92) accomplished in his 1759 paper *An Experimental Enquiry Concerning the Natural Powers of Water and Wind to Turn Mills and Other Machines Depending on Circular Motion.* While these experiments contributed to the long-term success of mechanics and hydraulics in the nineteenth century, they could not improve the design and settings of early modern waterwheels.

The fact that mathematics could not drive technological change did not prevent the rise of the waterwheel, which remained the best power source in many places well into the Industrial Revolution. Engineers were able to manage large-scale projects and solve problems in a practical manner that seemed to lay beyond the realm of feasibility. In 1685, for example, the French king Louis XIV had a new water plant built on the Seine. The Machine de Marly was designed to raise the water 150 meters and deliver it to the palace of Versailles to be used in the magnificent fountains. Fourteen large undershot waterwheels powered more than two hundred pumps. This technical achievement masks the fact that it was grossly oversized, resulting in a ruinous enterprise that was only successful as a prestige project (Reynolds 1983: 182–5).

TAMING THE SEA: DYKES, POLDERS, AND MAPS

In the early modern period, mathematics was increasingly used in large-scale territorial planning and improvement projects. Despite its considerable influence, the role of topography and surveying is often overlooked. The rise of nation states, bureaucratic administrations, and a better understanding of geography led rulers to enquire about and closely monitor their realms. The progress of practical geometry took a decisive step with the introduction and dissemination of surveying techniques based on triangulation. With simple instruments to record angles and distances, one could directly draw accurate maps using similar triangles drawn by intersections. In the Low Countries, Gemma Frisius (1508–55) first published the method in an appendix to Petrus Apianus' *Cosmographia* (1533). Its concrete use can be discerned in maps by Willem Hendrickz Croock or Jacob van Deventer (*c.* 1505–75) during the same period. Regional mapping projects quickly spread in the Low Countries as well as in Germany, where Matthias Öder (?–1614) began his survey of the Electorate of Saxony in the late sixteenth century.

Topographical mapping could be used to control one's realm or plan an attack on neighboring areas, but in itself this hardly counts as technological intervention in nature. Nevertheless, surveying information could serve as a basis for further land improvements. In the Low Countries, where Frisius produced a first sketch map of the region Brussels–Antwerp–Middelburgh using triangulation, people had been busy with land reclamation for centuries. Using dykes and draining techniques, many small swamps had been gradually turned into fertile land for agriculture, the so-called *polders*. These operations had begun in the Middle Ages, but took on a whole new dimension in the late sixteenth and early seventeenth century, based on more conscious planning and intensive use of surveying techniques by well-trained practitioners.

Polders are portions of low-laying land, mostly located below sea level, usually enclosed by dykes and drained in order to be made cultivable. They were an extremely sophisticated and complex technical construction for the early modern period, relying on sluices to manipulate the tides and on canals and windmills to constantly drain water. To get a sense of the intuitive use of geometry, one can read the *Tractaet van Dyckagie* (Treatise on Dykes), a manuscript written by the dyke master Andries Vierlingh (1507–79). The height of dykes had to be accurately measured, high enough to resist winter storms, and yet as low as possible to keep costs to a minimum without compromising safety. The slope of canals had to be carefully considered through leveling operations, while costly earthworks relied on volume calculation. Vierlingh made clear that mathematics was a keystone for polderization: "the most important is to have a good surveyor who can draw a map," he wrote emphasizing that this person "should have a good understanding of *geometria* and *arithmetica*" (Vierlingh 1920: 111, 276). However, his use of mathematics was still nascent, relying on unspecific instruments such as the plumb line, and giving lists of usual proportions rather than general formulas for the shape of dykes. Most notably, no reference to the new technique of triangulation can be found in his work.

The surveyor profession underwent a deep transformation in the Low Countries due to economic development, the Dutch War of Independence (1568–1648) and most importantly the many operations of land reclamation. To understand the role of a surveyor, let us follow the polderization of the Beemster Lake, north of Amsterdam. In 1607, a patent was given to a consortium of cities and merchants to proceed with the operation. Land surveyor Pieter Cornelisz Cort (–1608) was immediately asked to produce a map of the lake, using a surveying chain and a so-called "Dutch compass" (see Figure 6.2). Cort recorded not only the general shape of the lake, but also its depth and possible locations for future canals and windmills. This map was then engraved and sold by Willem Janszoon Blaeu (1571–1638), the Dutch cartographer famous for his globes and world atlases such as the *Theatrum Orbis Terrarum* (Theater of the World).

The draining itself began in 1609, but the ring dyke broke one year later due to terrible weather conditions. This proved beneficial for surveying operations: the following winter was so cold that the Beemster lake was completely frozen.

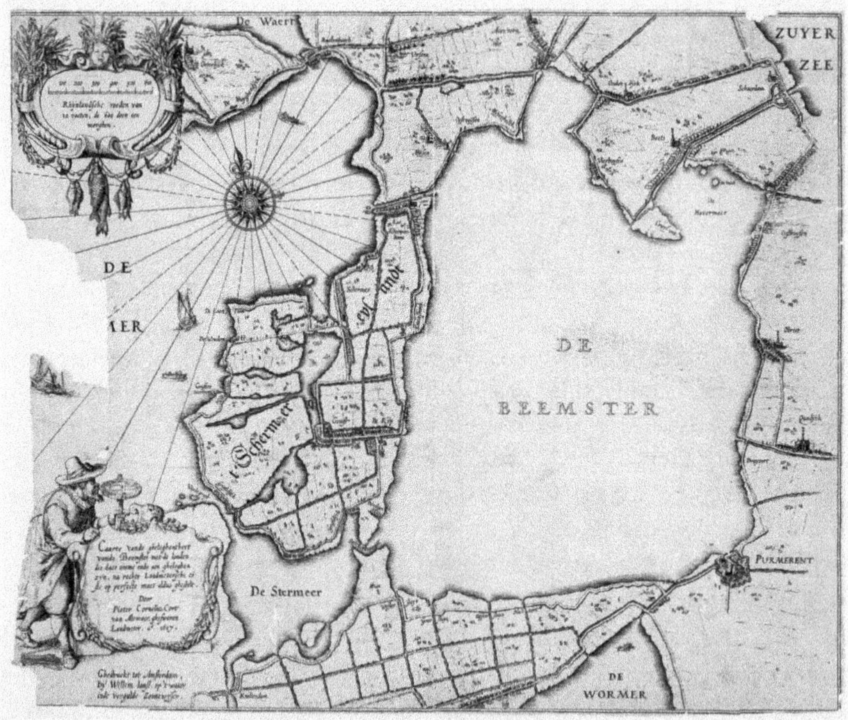

FIGURE 6.2. Map of the Beemster lake, 1607. Regionaal Archief Alkmaar: PR 1004144.

This allowed for a new survey made with chains, measured from a central point on the lake for greater precision. The allotment, that is, the division of the future polder into pieces to be sold to investors, could thus be made more precisely. Mathematics proved decisive in another respect: by bringing together land surveying and commercial arithmetic, long-term operations could be financed and legally secured.

The use of mathematics by Dutch surveyors and dyke masters allowed for the draining of dozens of lakes, thus deeply transforming Holland's landscape. The growing number of surveyors and their important responsibilities led to the introduction of formal training and examination, with the universities of Leiden and Franeker introducing engineering courses. The Leiden cursus was supervised by Simon Stevin, this was the so-called *Duytsche Mathematique*. In Franeker, Adriaan Metius (1571–1635), today mainly remembered for his study of π, was the professor in charge of mathematics, navigation, astronomy, and engineering.

The most successful textbook of the time was the *Practijck des Landmetens* (Practice of Surveying). It was written in 1600 by two mathematical practitioners, Johan Sems and Jan Pietersz Dou, who were successively engineers, surveyors, wine gaugers and cartographers. It was dedicated to Maurice of Orange, the *stadtholder* leading the revolt against Spain. Its content is both useful and ambitious: the authors neither use demonstrations nor follow the Euclidean pathway, but nevertheless expose decimal fractions in a very modern way, only fifteen years after Stevin's groundbreaking work *The Disme*. The *Practice* was completed by another booklet by Dou, describing the use of the Dutch circle in surveying operations (1612) (see Figure 6.2).

Such practitioners, whose skills and successes were widely recognized, were highly sought-after. Dutch surveyors and polder engineers were hired by many European rulers to reclaim and improve land. They worked in the cold swamps of Berlin and Danzig (Gdansk) in Prussia and in the more clement weather of Italian states. They were hired and poached by the kings of England and France, sometimes giving their name to the regions in which they settled such as the *Petite Flandres* (Little Flanders) in the southwest of France. Humphrey Bradley (?–*c.* 1625) was born in the Low Countries, but his father was an English merchant. After learning Dutch engineering techniques, he began working in England on the survey and draining of the Fenlands in Eastern England.

The draining of the English Fens would not be achieved in Bradley's lifetime, but would be pursued by Jonas Moore (1617–79), whose career epitomizes the success a mathematical practitioner could achieve (Willmoth 1993). Born into a modest family, he became known as a private teacher by publishing an *Arithmetick* followed by other popular textbooks. He was then appointed to survey and drain the Fens, producing a map of exceptional quality that established his standing as surveyor. He was consulted for many technical operations that required precise surveying, such as the fortification of Tangier in Morocco (then under English occupation), eventually becoming surveyor-general and member of the newly founded Royal Society of London. While few practitioners reached Moore's position, his example shows how technical successes could pave the way for scientific recognition.

SEEING THROUGH STONES

Another technical use of geometry in the early modern period was the *geometria subterranea* or underground surveying. This little-known mathematical "art" (*Kunst*), as it was labeled at the time, was developed in the metallic mines of Central Europe, where German-speaking miners extracted the silver and other precious metals that powered the nascent global economy. These surveyors and their expertise later spread to other mining regions, just as Dutch surveyors did in their own area of expertise, eventually reaching the rich mines operated by Spanish colonists in America and the Tsars in Russia.

Extracting silver involved many operations, machines and extensive knowledge, as literarily depicted in Georgius Agricola's famous book, *De Re Metallica* (1556). Several innovations, such as the introduction of elaborate waterwheels (see above) or the introduction of black powder allowed for deep-level mining that, in turn, led to complex geometrical problems. Mining shafts were narrow and sinuous tunnels which could be several hundred meters deep and kilometers long. It was crucial to know the exact geography of mines and to produce maps, not only in order to connect shafts, but also to avoid flooding and accidents. This type of surveying was much more difficult than the geometry of polders insofar as it had to be performed underground, in dangerous conditions where triangulation was in most cases unworkable.

The development of subterranean geometry follows a chronology roughly similar to its Dutch counterpart. In the first half of the sixteenth century, one notices the development of mostly empirical practices, where numbers and figures are only intermittently included in written reports. Early descriptions mostly rely on collecting specific problems and laying out rules of proportion. Mining maps began to be drawn in the 1530s, but practitioners still heavily relied on public measurements. When a dispute had to be settled, observations made with measuring cords in the mines of Tirol would be re-enacted in the open air, sometimes on frozen lakes, as sort of life-size maps.

Despite their seemingly rudimentary nature, these uses of mathematics were significant for at least two reasons. First of all, numbers and figures became everyday objects in the mining world: they were used for cost estimates of new machines, payouts of mining shares (*Kuxen*), for the shape of mining concessions and for the ceremonies of solemn measurement. Mathematics thus was a keystone of the mining states. Even sermons delivered to common miners in the (Protestant) churches emphasized the role of mathematics to set property limits:

> One should thank God for this, and praise the industrious people who achieve this with the triangle [. . .] This is why God gave [us] the liberal arts / so that one can recognize him / and thank him / and the fairness / and that the people are served by these / profitably and with good conscience." (Spangenberg 1574: 64–5, as quoted in Morel 2020: 28)

The main instruments of this mining geometry were the mining compass and the measuring chain, as well as the semi-circle and water level to measure the slopes of galleries. It was developed from a systematic use of similar triangles and more generally similar figures to produce maps and direct digging operations. Its outstanding development during the sixteenth century was temporarily stopped by the Thirty Years War (1618–48), in which silver mines were often besieged as strategic targets.

Around the middle of the seventeenth century, the introduction of a new, suspended mining compass led to a standardization of methods. Adapting trigonometric tables from Dutch mathematicians such as Simon Stevin or Adriaan Vlacq, subterranean surveyors were able to use their chain surveying to compute the horizontal and vertical components of a survey. They could then improve their mapping abilities with combined plane and elevation (cf. Figure 6.3). This can be seen in one of the few printed works on that topic, Nicolas Voigtel's *Geometria Subterranea* (Subterranean Geometry 1686), a work that was thoroughly studied by Gottfried Wilhelm Leibniz (1646–1716). The rarity of printed sources should not be mistaken for backwardness, but rather shows the ability of these state-controlled mining administrations to develop their own solutions. Surveyors were included in a grant system and trained both on the ground and underground, interacting with machinists and mining officials. They relied on handwritten textbooks that were gradually improved to adapt to new technical challenges.

This highly specialized training system produced skilled practitioners that worked all over Europe. They maintained close contacts with Scandinavian engineers working in the copper mines of Falun or in Christopher Pohlem's *laboratorium mechanicum* in Stockholm. German-speaking settlements were active in Eastern European mines, while individual surveyors had been recruited since the Renaissance in the Duchy of Lorraine (today France), in several French regions or in the Fugger mines in Spain. This example underlines the importance of surveying. Practical geometry did not only have a direct impact on the early modern landscape; it also enabled further technological developments in land reclamation and mining operations. These subdisciplines might well appear elementary from a theoretical point of view, but the specificities of local conditions and the need for cost-efficiency and practicability made it a highly specialized know-how. This mathematical knowledge was not only bookish or scholarly, but developed together with specific instruments, tables, and experiences.

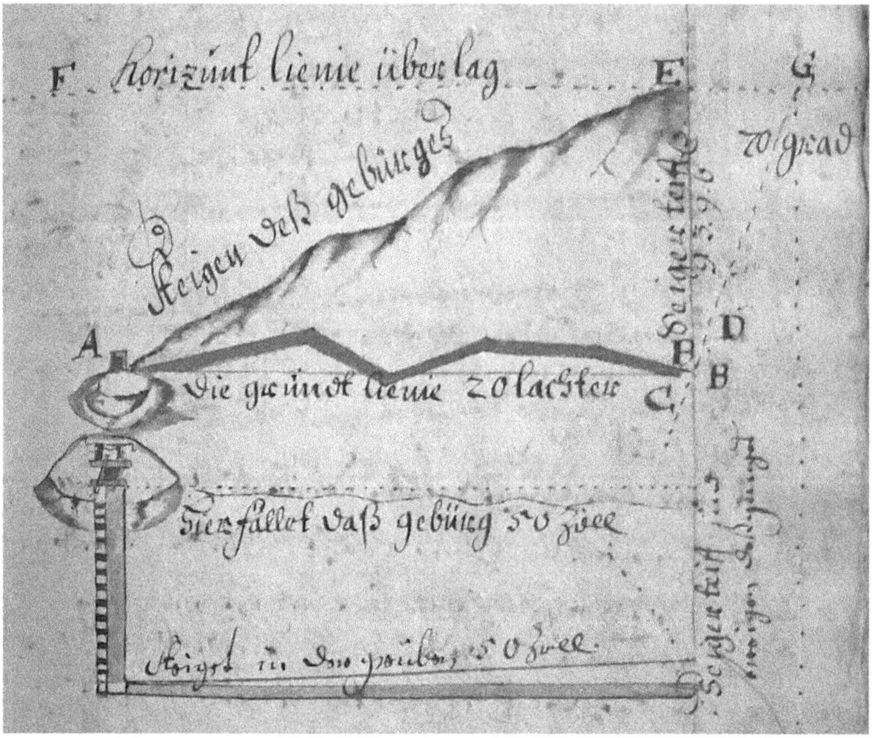

FIGURE 6.3. Plan of a Saxon mine by Johann Scherez, De Geometria Subterranea. Universitätsbibliothek Georg Agricola: WA XVII 15.

In the Dutch case, the geometry of polders was publicly traded through printed maps and textbooks, while German surveyors were privy to a more esoteric knowledge largely confined to the mining administration. In both cases, however, practitioners that are now all but forgotten were highly sought after in the sixteenth and seventeenth century, building wide and specialized networks that spanned not only Europe, but also extended to America and other European colonies. Compared to some other areas where mathematics was applied with mixed success, this ability to plan and monitor major geographical transformations was an impressive result.

PAPER TECHNOLOGIES: A PROLIFERATION OF TABLES AND CHARTS

In this section, we will take a look at a less obvious, yet crucial evolution of early modern practical mathematics: the growing use of numerical tables in all kinds of technological fields, from navigation to finance, and from gauging to gunnery. This growing use of tables is relevant for two reasons. First of all, printed or handwritten tables are a paper technology in itself, and its expansion over the considered time period is revealing. Second, the massive diffusion of tables allowed for a much broader use of numbers, a rise of quantification, and thus contributed to the general mathematization of civil life.

Useful sets of numbers could be found in any kind of business and used, though not computed, without formal training. English carpenters used to sell timber based on a "standard" log corresponding to a square base of twelve inches. Most logs, however, differed in their dimensions, and tables were thus produced to reduce various rectangular forms to this standard; one such table can be found in Leonard Digges' *A Booke Named Tectonicon* (1556). The availability of tables made it increasingly easy to deal with technical problems not only by relying on one's experience or know-how, but by combining these with a highly efficient way of information storage. This allowed for faster and more reliable information as well as for standardized procedures.

This evolution was itself based on two innovations. The technology of papermaking developed in China in the second century and was adopted in the Arabic world during the eighth century, but only spread to most European countries in the fourteenth, where it was quickly produced using waterpowered paper mills. A few generations later, the invention of the printing press with movable metal types by Johannes Gutenberg (*c.* 1400–68) made it possible to replace handwritten manuscripts with cheaper solutions. It gradually became common to store and circulate numerical tables with much less mistakes than those made by copyists at a fraction of the cost, making them appealing for merchants and technicians.

Scholars and mathematicians made great use of this opportunity, publishing everything from prime tables (tables of prime numbers) to tables of square and cubic numbers. The method of logarithms was introduced to speed up computations by John Napier (1550–1617) and improved tables were soon published by Henry Briggs (1561–1630) and Adrian Vlaacq. These tables, used together with the increasingly common trigonometric tables helped hasten tedious astronomic computations, for example as used by Johannes Kepler (1571–1630) to produce his famous *Rudolphine Tables*.

Outside of scholarly circles, printed and handwritten numerical tables developed just as quickly. Finance can be used as a first case study: the calculation of interest, and above all of compound interest, required a lot of computation. In the late Middle Ages, interest tables, which could be used to calculate compound interest over several quarters or years, and later to set the amortization of loans, were not published. Such a table can be found in a fourteenth-century manuscript entitled *Practica della mercatura* (Practice of Commerce) by the Florentine banker Francesco Balducci Pegolotti.

This quickly developed in the following century despite a theoretical ban on usury in Catholic countries. Johannes Widmann (*c.* 1460–?) printed his *Behende vnd hubsche Rechenung* (Mercantile Arithmetic) in 1489, a work in which he introduced the modern symbols for plus and minus signs: "+" and "−". He computed the value of a loan of thousand florins after four years, provided that the yearly interest rate was 4 percent. In his *Arithmétique* (1558), the French mathematician Jean Trenchant gave several examples of compound interest used in merchants' calculations. While it is sometimes difficult to assess if such books really dealt with useful knowledge or only with complicated scholarship, Trenchant went on to present a new method:

> Nowadays, merchant bankers lend money to princes . . . a method was recently invented during All Saints' Day fair 1555, so that for every 100 pounds, the king would pay them 5 at each fair: but after 41 fairs or payments, he would be discharged of the principal and everything. (Trenchant 1558: 243–4)

This refers to a real event, for King Henri II of France had indeed used this very method for his *Grand Parti* in Lyon in 1555, refinancing his public debt and borrowing money at the same time. With this new technique, it was possible to pay interest regularly while gradually amortizing the principal—a now common feature in many loans. Although this brilliant innovation was mathematically sound, things turned sour when the king went bankrupt a few years later, ruining his lenders. While compound interest was not unknown, as previous examples have shown, this knowledge was not widely circulated and innovations were mostly limited to important merchants, financiers, and kings. In the sixteenth century, the Low Countries and their main trading center, Antwerp, overtook the Italian cities to become the center of European finance. In this context, Simon Stevin (1548–1620) built on Trenchant's work and published the first *Tafelen van Interest* (Tables of Interests) in 1582.

It was the first time that a book dealt not merely with a set of examples but gave both a coherent theory of interests and a full set of tables that could (almost) immediately be used by anyone who knew how to perform a rule of three. Stevin's work was reprinted, translated into French and, most of all, widely imitated. These new technologies—lending with compound interest and computing the real cost of moneylending—would develop in the following decades. Financial innovations then played a great role in transforming European economies, conferring the Low Countries and England, the two most advanced countries, great advantages over their rivals.

Another example of paper technologies are the instruments and tables used to gauge casks and barrels. With the rise of merchant capitalism, European goods were increasingly circulated using ships. Barrels were the most common container, used to bring wine from Porto and Bordeaux to England, to ship grain from Sicily to Northern Europe, and more generally to transport goods as diverse as molasses or fish. Computing the exact volume of a barrel, or the volume of liquid in a partially filled barrel, was a matter of importance in order to properly sell and tax its contents. Moreover, due to the shape of most barrels, it was an extremely difficult problem that would only be theoretically solved in the eighteenth century, after the development of integral calculus.

Despite its difficulty, this problem had to be tackled due to its fundamental economic importance. In the early modern period, practitioners and scholars had to rely on approximation techniques. It is, for example, possible to see a barrel as a pseudo-cylinder and thus use the corresponding formula, obtaining its volume by multiplying the circular base by the length. A more precise approximation could be obtained by seeing a barrel as two truncated cones connected at their bases. Even these simplifications required a non-trivial knowledge of mathematics. One had to master some practical geometry to compute volumes (this was named *stereometry*) and possess a good approximation for π, such as $\frac{22}{7}$. In practice, however, even a well-trained gauger could not take the time to measure each and every barrel in such a way while working in a busy marketplace.

A method was needed to quickly produce an estimate that would be acceptable by all parties, and it required a gauging rod (*virga visoria*). It developed in the south German merchant cities such as Nuremberg or Frankfurt-am-Main, located at the crossroads of major mercantile lines, from the Low Countries to Hungary and from France to Bohemia. The first instruction on how to build a rod was found at the turn of the fifteenth century in the *Chronicle* of a Nuremberg patrician, who also happened to possess the first paper mill in town—an illustration of how technical progresses arose simultaneously in the context of urban mercantile classes.

The most common gauging rod, the so-called square-rod (*Quadratrute*), considered the barrel as a cylinder. One would thus measure its length and the diameter; since the object was no perfect cylinder, the gauger would simply try to find the medium diameter to limit errors. Since the volume of the cylinder is proportional to the length and to the square of the diameter, the rod would be graduated using a linear scale for length on one side and a square root scale for diameters on another side (cf. Figure 6.4). A professional gauger could thus quickly take both measures, and the only needed computation was multiplication, for which all textbooks provided a table. Reckoning masters (*Rechenmeister*) began to teach these skills and to include them in their syllabus. Dozens of such books were written in the sixteenth and seventeenth centuries, always in vernacular, and contributed to the circulation of this practical arithmetical knowledge.

Gauging rods were instruments that embodied these mathematical tables. A gauger would use them as one uses an algorithm: instead of learning the formula giving the volume of a cylinder and performing complex calculations, he would collect two pieces of data in these tables and simply multiply them to obtain his result. Needless to say, such a method could not produce an exact result.

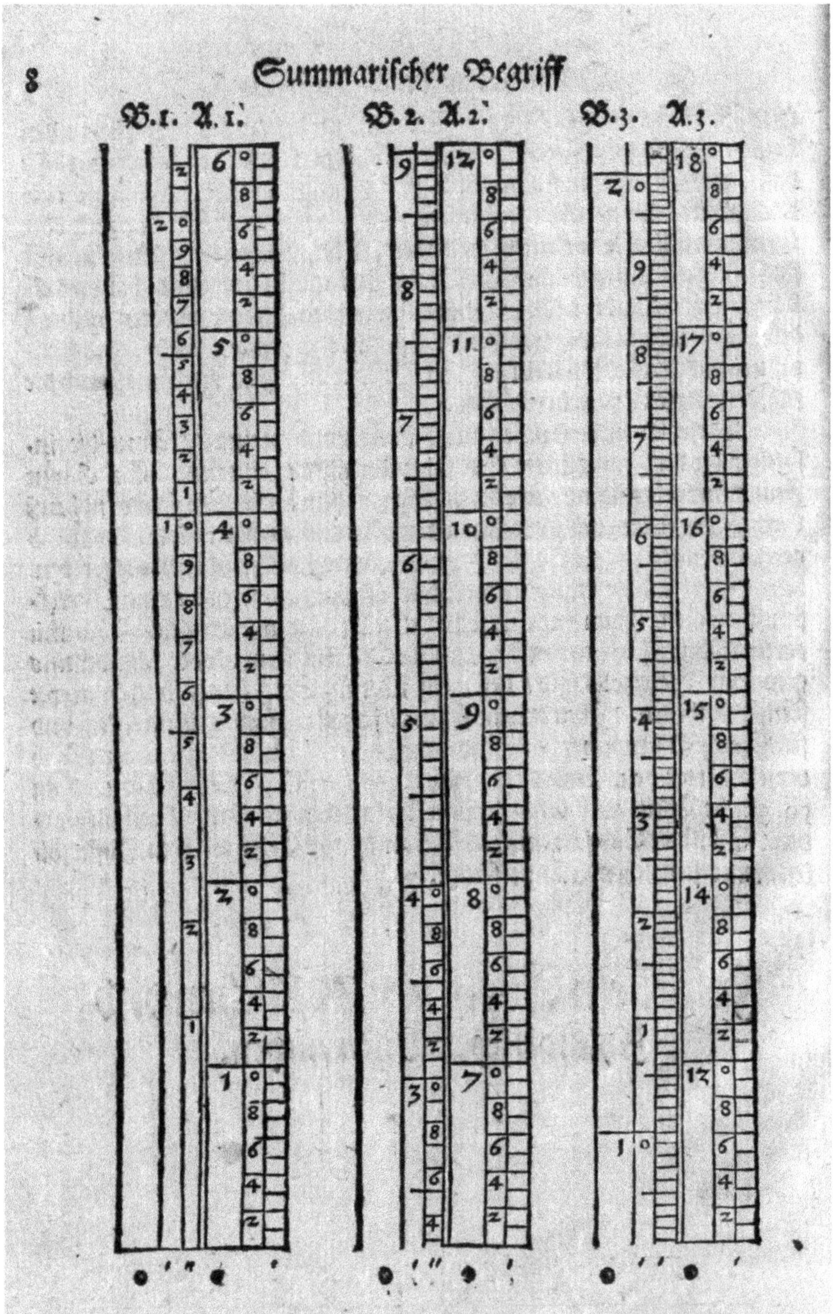

FIGURE 6.4. Drawing of a gauging rod. ETH-Bibliothek Zürich, Rar 583.

However, all that mattered in practice was that gauging rods could quickly produce an estimate precise and consistent enough to be socially acceptable.

A further development of the gauging rod was the cubic rod, or *Kubikruten*. This rod was gauged using a cubic scale, and allowed the determination of the volume of a barrel using a single measurement. One should insert the rod through the bunghole of a cask and reach to the furthest possible point, as can be seen on Figure 6.5. Each cubic rod assumed a particular relation between the length and depth of the cask, thus giving a precise estimate only for a specific type of barrel. This in turn might have led to a standardization of cask-building: in that case, early modern mathematical practices and technological standardization went hand in hand.

As can be gathered from these two examples, mathematical paper tools were a booming technology in the early modern period. While they could take various forms such as plain tables, charts, diagrams, or even rods, they ultimately relied on the same principle. Series of numbers were computed and then incorporated in a concrete form, allowing merchants, artisans and craftsmen to reach the desired results mostly with manipulation and simple operations. Saxon bakers relied on "bread tables" (*Brotordnungen*) to ascertain the price and weights of breads, English carpenters obtained the value of a log from "tables of Tymber," while navigators from all countries used simplified trigonometric tables to ascertain their position (cf. Chapter 5). Thousands of mathematical practitioners, be they bankers, sailors, or gaugers, learned to use these paper technologies incorporating elaborated mathematical methods.

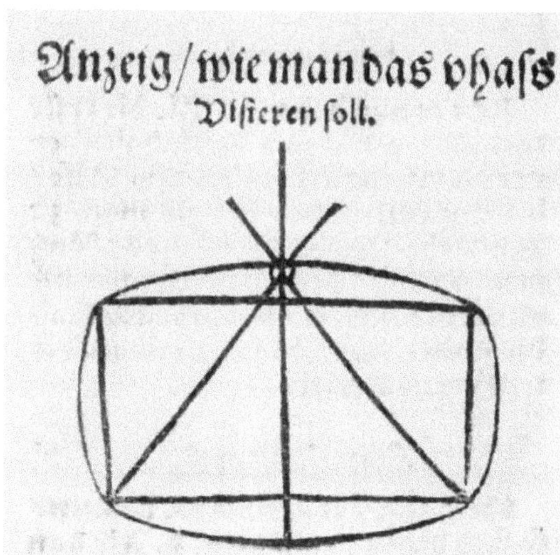

FIGURE 6.5. How to measure a cask. ETH-Bibliothek Zürich, Rar 583.

Although they could be used in a purely intuitive way without understanding the full underlying theory, their use reflected a growing numeracy outside of scholarly circles and in everyday life.

THE RISE OF GEOMETRIC PLANNING AND ARCHITECTURE

Uses of mathematics in early modern architecture are more difficult to assess. While the general trend likely implied a growing use of numerical information in the seventeenth century, many artisans primarily relied on their experience. For individual and small-scale projects, there were "no engineering calculations based on mathematically formulated laws of nature"; similar to today, knowledge accumulated by trial and error was reliable enough for generic buildings (Knobloch 1994: 251).

It would be inexact to claim that no literature existed on architecture. After all, the rediscovery of ancient works had led to new editions of the Greek Archimedes (third century BCE) and the Roman Vitruvius (first century BCE). Both had written about architecture, theorized about simple machines such as the lever and the screw and proposed more complex machines to solve difficult technical problems; however, "these writings had no influence on the practice of medieval craftsmen who did not understand Latin" (Knobloch 1994: 251).

The situation was markedly different for large-scale projects. The early modern period saw the growing influence of states, but also of administrations, chartered companies, and venture capitalists. These new entities required mathematics for several reasons. Prestigious projects were launched to show the power of a city or its ruler; these political statements often entailed new technical challenges that no existing methods would be able to overcome. Artist-engineers would thus be needed to solve these problems in a harmonious way. The rapid transformations of nascent colonial states, constantly at war, also created new architecture for major infrastructures such as ports, canals and forts. Mathematics could be used both as geometry, to rationalize the use, and as arithmetic, to limit the costs.

Giulio Savorgnan (1510–95) illustrates both trends. As engineer and *capo di guerra* for the Republic of Venice, he oversaw the rebuilding of the fortress of Nicosia, the largest city of the Cyprus Island. This fortification is remembered for being a rare example of a perfect shape of a hendecagon, an eleven-sided polygon (cf. Figure 6.6). It was considered perfect not only for its geometrical shape, but also for the rare ability of the engineer to respect both the timescale and the budget—eminent matters of practical arithmetic.

In this section, three emblematic projects spanning three centuries will be discussed. The first is the construction of the Dome of Florence by Filippo Brunelleschi (1377–1446). The second is the transportation and re-erection of the Vatican obelisk in Rome by Domenico Fontana (1546–1607). The third is

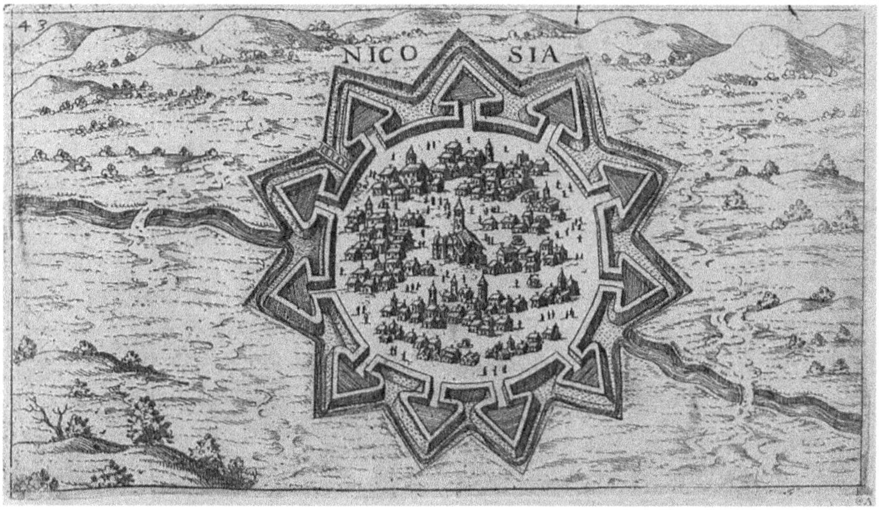

FIGURE 6.6. The fortress of Nicosia in Giuseppe Rosaccio, *Viaggio da Venetia, a Costantinopoli per mare, e per terra*, Venice, Giacomo Franco, 1598. Courtesy of Laskaridis Foundation / Europeana.

the building of the *Canal du midi*, joining the Atlantic Ocean to the Mediterranean Sea, under supervision of Pierre-Paul Ricquet (*c*. 1609–80).

The city of Florence, in the Italian province of Tuscany, was a powerful early modern state where the aristocracy, most famously the Medici family, supported arts and crafts. The building of a new cathedral, Santa Maria del Fiore, began in the late thirteenth century. The huge enterprise was almost finished at the beginning of the fifteenth century, when only the dome was left to be built. A competition was launched in 1418 for this project. Large domes had already been built by the Romans, most notably for the Pantheon in Rome. But the recipe for Roman concrete had been lost, making these achievements impossible to replicate.

Since no comparable method existed, a new technical solution had to be worked out. The main challenge was that the new dome was so big that it had to be constructed without a temporary wooden frame to support it before completion, and thus needed to be self-supporting. A brick model of several meters built at the beginning had been conserved within the cathedral: one had to scale the model to a real-size dome that would support itself. Brunelleschi produced a document, scarce on technical details, explaining his intention to erect a dome with a double shell. He produced his own wooden-and-brick model and used it repeatedly to design methods during the work. A fifteenth-century engineer was not only responsible for the geometrical design, but controlled all aspects of the construction; Brunelleschi even designed new machines to lift the materials. The dome was inaugurated by the Pope in 1436. The total height of the building, completed long after Brunelleschi's death, would be 110 metres, with a dome made of several million bricks and weighting about 25,000 tons.

Little is known about the exact use of mathematics by Brunelleschi, who left almost no traces of his work. The design was likely not produced theoretically, but Brunelleschi relied on his long experience with the physical properties of materials. Trained as a goldsmith and an architect, he also had a great interest in painting, drawing and proportions. In his biography, written by Antonio Manetti (1423–97), Brunelleschi's studies of ancient buildings proportions together with Donatello (*c.* 1386–1466) are recollected:

> In many places they had excavations made in order to see the junctures of the membering of the buildings and their type – whether square, polygonal, completely round, oval, or whatever. When possible they estimated the heights [by measuring] from base to base for the height and similarly [they estimated the heights of] the entablatures and roofs from the foundations. They drew the elevations on strips of parchments graphs with numbers and symbols which Filippo alone understood. (Manetti, Life of Brunelleschi: 52–3, quoted in Long 2011: 67)

Brunelleschi could then use these data sets, together with his practical experience, in order to propose new sets of proportions and try them out on models. These were likely used to test the resistance of the shells: although scaling problems do not allow for a direct computation, the results could be skillfully transferred by an experienced engineer. A newer use of these models, which could be several meters large, was the use as a control tool, to check if the distances between elements under construction were what they should be.

Mathematically, this method is valid, since it only rests on proportionality. According to the historian Salvatore di Pasquale, the tool used to measure distances on the model was the *diffinitore*. This mathematical instrument could record angles and distances between two points and was described by Leon Battista Alberti in his *De statua* (On Sculpture, published 1464) as a tool to reproduce statues with different scales. Alberti also produced the groundbreaking Renaissance book on construction engineering, the *De re aedificatori* (On the Art of Building, printed 1485, written 1443–52).

In this respect, Brunelleschi's method was new in its rigor: "he worked with ground plans and elevations drawn to scale on a mathematical basis" (Krautheimer 1970: 240). He exerted a tremendous influence on later Italian artists-engineers. In the late fifteenth century, it was common for architects to visit prestigious buildings and draw sketches to capture both the general proportions and specific details. Innovations proposed by Brunelleschi found their way into the notebooks of Buonaccorso Ghiberti (1451–1516) and several other architects of the time. Historian Elizabeth Merrill underscores that fifteenth-century artists-engineers shared the "same mathematically-based design practice" (Merrill 2017: 27). They could act as civil architects, but also knew how to design fortifications or build various kinds of machines.

Several features of the use of early modern mathematics for architecture can be derived from this brief sketch. First of all, it mostly concerned exceptional projects and problems for which no existing solution could be found. Interestingly, there was a form of publicity around these enterprises: while Brunelleschi hardly wrote down technical details, his methods were nevertheless studied, reproduced, and empirically improved. Following generations would pursue his studies, as both Leonardo da Vinci and Francesco di Giorgio did to achieve the Dome of Milan in the 1490s. Moreover, such constructions were increasingly being awarded during official competitions where projects were evaluated. To be awarded the contract, Brunelleschi had to prove both support from the Medici and offer a technically viable proposal, to be examined by a college of experts. Technical and scientific qualities were not the only criteria—politics played an important role, just as it does today—but this means that engineering was gradually acquiring a public character.

Construction projects were neither directed by scholars nor produced according to mathematical theories. In contrast, the artist-engineers that were involved possessed a vast knowledge of ancient architecture and used models, measurements, and precise numbered information. The scale and duration of such projects involved many people, ensuring at least a limited circulation of knowledge. Spanning over several generations and involving many different trades, it also ensured the diffusion of skills.

Our second example, the Vatican obelisk had been brought by Caligula to Rome from Egypt in the first century BCE, where it was supposed to have witnessed the crucifixion of St. Peter. When sixteenth-century popes decided to replace the old St. Peter's Basilica, built in the fourth century, it was decided to move this important relic on the square of the new St. Peter's Basilica. After eighty years of efforts, the new Pope Sixtus V appointed Domenico Fontana (1546–1607) to help build the dome and move the obelisk, after a committee had evaluated several propositions.

While the Florentine dome (see above) could be seen as a model for the new Roman dome, the task of moving and raising a stone weighing 327 tons was a new challenge. In the spirit of the Renaissance, trying to match the achievements of the antiquity would lead to a modern and essentially new solution to an age-old problem. Although it was only moved about 200 metres, the transport and subsequent erection of the Vatican obelisk in 1586 is remembered today as a milestone in the history of engineering. This is mostly due to the efficient dissemination of information about this event: in 1590, Fontana published a richly illustrated book on this topic, *Della trasportatione dell'obelisco vaticano* (On the transportation of the Vatican Obelisk).

One could see how the obelisk had been enveloped in a wooden structure he called *castello* (castle), lifted with complex equipment and numerous ropes and put on a special carriage. It was subsequently lowered, and later moved horizontally to its destination. Hundreds of men and about seventy-five horses then activated pulleys to raise the obelisk. In his book, Fontana described that he won the contest by producing a working model before the committee. In this work, he would reproduce not only his model, but added seven other unsuccessful proposals, as can be seen in Figure 6.7.

Fontana's book, as well as another *Liber Mechanicorum* (Book on Mechanics) by Filippo Pigafetta (1533–1604), and a dozen minor texts published in the 1580s, explain at length the theory behind the operation. Elementary Euclidean geometry was involved to measure the obelisk. Its weight was then computed by weighing a sample of the material and inferring to the whole volume of the

FIGURE 6.7. Seven discarded machine designs for moving the obelisk, from Domenico Fontana, *Della trasportatione dell'obelisco vaticano*, 1590, p. 8. Bibliothèque de Institut national d'histoire de l'art: NUM FOL RES 126.

monument. The theory of mechanics (statics) could then determine the power needed to raise it. Having proven his abilities, Fontana was then hired to move other Roman obelisks, although these operations attracted far less publicity.

However, it is good to remember that the operation also represents the development of engineering in the sixteenth century. Not only was Fontana an architect involved in many construction sites, he also collaborated with Ignazio Danti (1536–86) to move the obelisk. A priest and cosmographer, Danti produced numerous maps, globes, and instruments. He published treatises and taught mathematics at the University of Bologna. He was asked by the papacy to assist Fontana, but also to work on the reform of the calendar. One thus notices the extension mathematics was gaining in the sixteenth century. Artist-engineers, architects, and professors could be consulted in many instances, in which their mathematical skills were seen as a multifarious tool for practical matters.

The last example of a major engineering project is the building of a canal linking the Atlantic Ocean to the Mediterranean Sea in the southwest of France. The project itself was anything but new. When King François I brought Leonardo da Vinci to France in 1516, he had already hoped to apply the techniques seen in Italy to connect the Garonne River, which flows into the Atlantic, with the Aude River, flowing into the Mediterranean. In the late sixteenth century, dyke master Humphrey Bradley (see above) had proposed a similar solution.

These proposals had never been funded, given the difficulty and cost of the enterprise: the final canal would be more than 200 kilometers long. The most difficult aspect, already mentioned by Leonardo, was the divide between the two watersheds, almost 200 meters in height. This implied the need not only for hundreds of locks to elevate water, but also for an important water supply precisely at this highest point; otherwise, the canal would ineluctably empty itself in the two seas.

The canal was finally dug as a prestige project under the reign of the Sun King Louis XIV. It was part of the administrative reforms brought on by his Controller-General Colbert to ensure a firmer grip on the southern part of the country. A proposal made by the then little-known Pierre-Paul Ricquet (1609–80), a tax collector in Languedoc, was accepted, in the hope that commercial and military benefits of circulating ships directly through France, avoiding rival European powers, would make up for the huge costs.

The scale of the project and the new technical challenges that had to be met are major differences between the *Canal du Midi* and the previous examples of Brunelleschi and Fontana. Both Italian engineers had managed to come up with modern solutions to replicate ancient achievements in dome building and obelisk transportation. Nothing remotely similar had ever existed, to the point that a recent historical study on the canal was entitled *Impossible Engineering* (Mukerji 2009). While the adjective might be slightly far-fetched, building such a canal required the successful handling of two sets of issues. The first one was logistical:

one had to coordinate thousands of workers and dozens of engineers, to subcontract hundreds of engineering structures and mutually dependent problems. The second set of issues was that each of the engineering problems involved advanced practical mathematics: one had to design new locks, to come up with reservoirs of great capacity, to design a first canal (named *rigole*) to bring enough water from neighboring mountains to supply the canal at its highest point, to compute the water flows, to estimate losses due to evaporation, and so on. If this project was successfully carried out in the seventeenth century, it is precisely because the nation states of Western Europe had developed into powerful entities that could afford and sustain such ambitions, both economically and politically.

Unlike Brunelleschi, a skilled architect who personally oversaw all construction work and kept his methods secrets, Paul Ricquet acted as coordinator of the work, reporting to the Controller-General Colbert in order to sustain the financing of the project. Most of the engineering work was carried out by others, foremost by François Andréossy (1633–88), an engineer trained in Paris and Italy who "calculated water flows from mountain sources, measured the lengths of each section of rigole, drew maps that included topographical features relevant to the canal and worked on technical specifications" (Mukerji 2009: 48). His work on leveling and mapping relied in turn on measurements made by several surveyors whose last names have often been forgotten. After producing a *plan géométrique* in 1666 to show the feasibility of water supply, Ricquet nevertheless had to demonstrate his idea by digging a small-scale channel. Progress in surveying and mapping methods had increased their importance, but maps and models had yet to replace the convincing power of real-life experiences.

When digging operations began in 1667, one year after the contract had been awarded to Riquet, Colbert and the French administration sought to control it as tightly as possible. The team of engineers led by Ricquet had to report to a commission led by Louis Nicolas de Clerville (1610–77), military engineer and *commissaire général des fortifications* and to reach technical compromises. The structure of the project reflects the development of seventeenth-century administration: a local entrepreneur worked under the control of experts from the capital, and their common language was the mathematics used for quotes, contracts, and plans.

Nevertheless, one should not overstate the extent to which mathematics was used in this project. While elevations and leveling were certainly accurate, the domains of hydraulic and material mechanics were still prone to errors. Locks proved to be the weak point of the canal in several instances. Despite the experience gained in Italy and in the Low Countries, local considerations made adaptations necessary, and no scholar of the time was able to do this mathematically. When it came down to it, experience often had the upper hand over theory. In other words, "precedents rather than principles were used for the design." The true innovation, necessary to build staircases of locks, was the introduction of curved locks; but there, as with the computation of the proportions of locks, it was in most cases local artisans who proposed functioning solutions, not engineers or university professors.

To sum up, the *Canal du Midi* is a powerful symbol for the use of mathematics in the early modern period. It perfectly combines both the significant progress of mathematization and its numerous shortcomings. Engineers and mathematical practitioners were increasingly able to tackle technical problems using their knowledge and know-how, but new solutions were rarely worked out perfectly through modelization on paper, as would increasingly be the case in the following centuries.

The legacy of the *Canal du Midi* illustrates this ambiguous efficiency of mathematics. Although it was completed in 1681, one year after Ricquet's death, it had to be improved by the famous engineer Vauban (1633–1707) in the late seventeenth century in order to be fully operational. Even so, it never fulfilled its original goal: its dimensions and the water flow were not sufficient to carry the French navy from one sea to the other, or even large merchant ships. It was actively used mainly by locals, opening up the secluded *Hinterland* of southern France. Further extensions were never carried out, for the high maintenance costs made the canal barely profitable.

CONCLUSION

In this chapter, university mathematics and scholars have been mentioned mainly in passing, leaving most of the space to engineers and concrete projects. This does not mean that the world of techniques and theory never met, but rather that achievements were foremost the work of practitioners. Acute observers such as Francis Bacon (1561–1626) could promote empiricism and observation, but their intellectual influence had little immediate impact and would only translate into actions in the long term. Mathematics still occupied a relatively low position in European universities, and could be neglected in academic circles as being "the business of Traders, Merchants, Seamen, Carpenters, Surveyors of Lands and the like" (quoted by Bennett 1991: 189). The French humanist Peter Ramus (1515–72), in contrast, urged his audience and readership to reach out to practitioners. He saw geometry not as the science of certainty or demonstration, but as the art of "measuring well," stating in 1569:

> This aim appears with more splendour in practice and in geometrical works than in precepts and rules, as we see astronomers, geographers, surveyors, navigators, mechanics, architects, painters and carvers using nothing but geometry for the description and measuring of stars, lands, machines, seas, buildings and paintings. (Hooykaas 1958: 25)

This set of examples offers a way to reconcile diverging accounts about the efficiency of early modern mathematics used in technical fields. As testimonies of their contemporaries show, achievements of artists-engineers were greeted with optimism and enthusiasm. In several areas such as surveying or mapmaking, successes were incontestable. On a broader level, an expanding low-key mathematization led to growing social recognition of mathematical practitioners. The early modern period witnessed countless steps towards mathematization in many domains. Historiography tends to focus on the huge enterprises and demonstrations of prowess such as the Dome of Florence or the waterworks of the Versailles Castle. These do indeed deserve our awe, but mathematics was much more pervasive in daily techniques. This ongoing mathematization of civil life was uncoordinated and mostly inarticulate, with the possible exception of the Low Countries. Still, it was becoming increasingly natural to rely on arithmetic and geometry in planning enterprises, to ensure trust and accuracy, even in legal arguments.

Outside academies and universities, countless forgotten practitioners were transforming the European landscape by designing new machines, building polders and digging mines. Crafts increasingly relied on mathematics and most skilled jobs came to include tables, charts, and various simple instruments. These tools are not scientific tools per se, but their daily use relied on basic skills in mathematics, reflecting a growing numeracy. Overall progress, however, was so uneven and modest as to be hardly recognizable by contemporaries. In many domains, engineers and artisans faced not only the complexity of nature, but also the vicissitudes of human affairs. To give but one example, attempts to unify measuring units, an important precondition for the standardized technology of the Industrial Revolution, would fail in all political settings, from the Republic of the Low Countries to the absolute monarchy of France.

From our modern point of view, early modern efforts can appear uncoordinated, sporadic and sometimes unsuccessful. They undoubtedly lacked both the coherence and explicative power of science-based modern engineering. Important areas were left untouched and important attempts failed, as the case of waterwheels reminds us. However, one should look beyond the strict distinction between a fully successful application of science and a "craft practice confined to rules of thumb handed down from one generation to the next" (Cohen 2010: 309). To a close observer such as Ramus, who could not foresee the place that mathematics and science-based engineering would gain in the future, these attempts, however modest and uneven, were already impressive.

CHAPTER SEVEN

Representing Mathematics

Promoting the Mathematical Sciences in the Early Modern Period

ROBERT GOULDING AND VOLKER REMMERT

In the modern age, in which the humanities struggle to be noticed alongside "STEM" (science, technology, engineering, and mathematics) subjects, it can be difficult to appreciate how irrelevant and obscure mathematics seemed for many early modern people. In Europe, many humanists dismissed it as a barbarous relic from the medieval universities. While the celebrated humanist Erasmus of Rotterdam (1466–1536) could see some (though not very much) interest in geometry, he advised students to take no more than a "taste" of the arithmetic and geometry that underpinned it; such studies were incompatible with a genuine interest in human beings and their cultural productions (*De pronunciatione*). At the Renaissance universities, ordinary students of the liberal arts, whose future depended on their dialectical skills above all else, found it an irksome irrelevancy (Goulding 2010: 86f.). Arithmetic and geometry had practical uses, of course, in surveying, navigation, and trade, which by the way did not endear them to Renaissance intellectuals. Such arts of calculation bore little resemblance to the advanced mathematical studies made possible by the recovery of Greek mathematical texts—a recovery that was firmly rooted in the humanist enterprise—and the mathematical innovations of the sixteenth and seventeenth centuries (cf. Chapter 3). From the research of Sonja Brentjes, it seems that the situation of the mathematical sciences was similar in Islamicate societies (Brentjes 2018). Nevertheless, the mathematical sciences found their supporters, both within and beyond the academy, and they sought recognition both for themselves and for their arts.

We will survey some of the means by which early modern mathematicians represented the mathematical sciences, particularly to those who were predisposed to find little interest in the subject: humanists, princes and other government officials, and their own students. As the study of history gained status and earned a place in the early modern academy, recounting the history of the mathematical sciences provided an opportune vehicle for piquing student interest in a subject that might otherwise have offered little interest. Mathematical knowledge could thus be brought within the liberal arts and could even contribute to their study.

Yet, the humanistic purpose of the mathematical sciences was not universally accepted. Some argued that the best mathematical education consisted in practice and imitation of artisanal, practical, or applicable branches of the mathematical sciences such as surveying or bookkeeping. One of the most striking ways to illustrate the utility of the mathematical sciences was to extol the almost unbelievable marvels that could be brought about by the skilled practitioner (cf. Chapter 6).

Such representations of the mathematical sciences were laid out in prefaces to textbooks and proems to lectures and orations. But they also naturally lent themselves to visual representation, whether as the frontispiece to a printed book, in paintings or frescos, or by adorning mathematical instruments. Such visual representations of the mathematical sciences, their history and cultural significance, could recommend their study and patronage in a courtly context, especially to a ruler who might appreciate artistic wit and beauty over arguments about the nature of mathematical reasoning. Nevertheless, when potential patrons were courted with written depictions of the mathematical sciences, these texts were more likely to emphasize their utility in war or statecraft, than their antiquity or liberal nature.

HISTORIES OF MATHEMATICAL SCIENCES

In the academy, lecturers in the mathematical sciences had to cover the curriculum inherited from the medieval university—based on the *quadrivium* consisting of arithmetic, geometry, astronomy, and music. Often, they attempted to gain their students' interest by connecting the mathematical sciences to their other studies and constructing a place for them among their more humanistic interests. Ancient sources underlined the great antiquity of the mathematical sciences which, among humanists, was almost in itself an advertisement for their worth. In his *Antiquities of the Jews*, the scholar Flavius Josephus (*c.* 37/38–100), whose works were widely read in the Renaissance, had traced the origins of all human knowledge back to Garden of Eden. It was then transmitted through the Flood by inscriptions on pillars, one of brick and the other of stone, and fashioning the authors Adam and his son Seth as biblical fathers of arithmetic, astronomy, and geometry (see Figure 7.1).

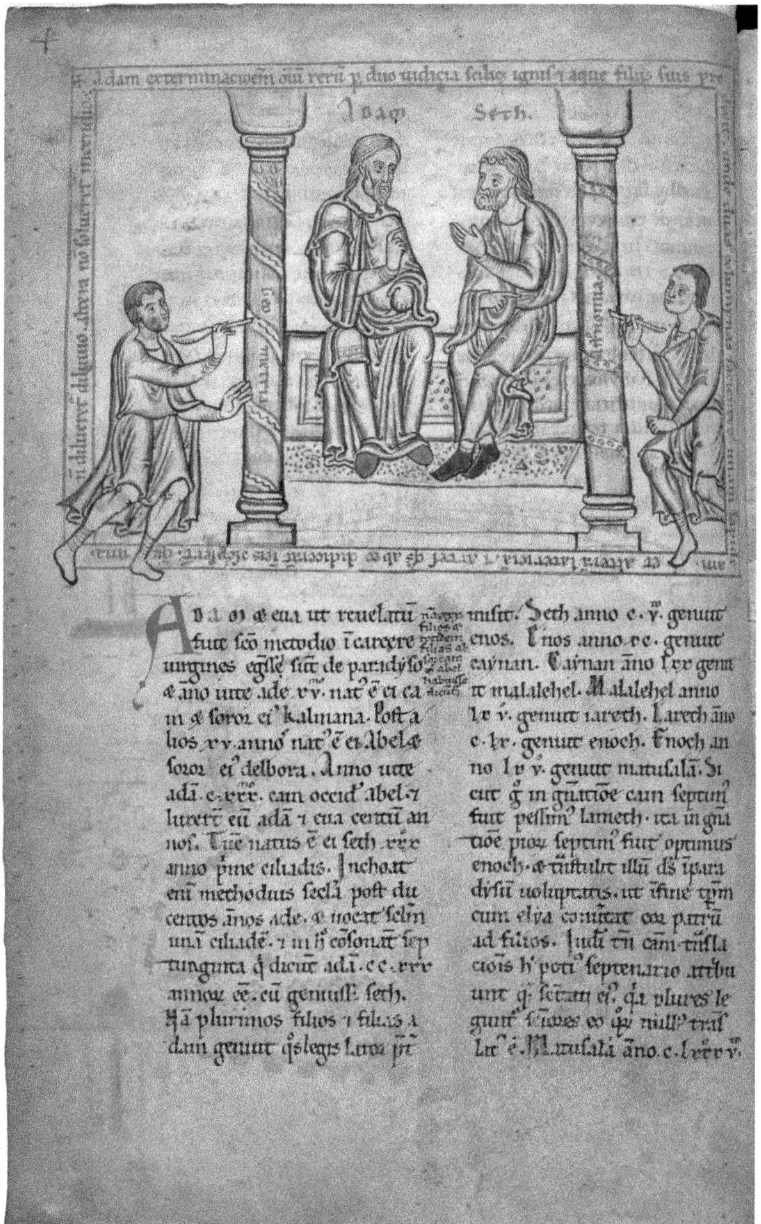

FIGURE 7.1. Adam and Seth, in: *Chronica regia coloniensis*. Herzog August Bibliothek Wolfenbüttel: GS 04-0515.

As Renaissance universities included humanistic studies such as history in their curricula, alongside philosophy and the liberal arts, educators revived a pedagogical practice that itself had a long historical tradition: teaching the mathematical sciences within a historical context. One of the earliest extant examples was that of the Neoplatonic philosopher Proclus (412–85), who initiated young students at his Athenian school into the study of metaphysics through the study of the first book of Euclid's *Elements*. These lectures, later published as a commentary on the same mathematical text, drew upon the *History of Geometry* by Eudemus of Rhodes (fourth century BCE) to frame the history of mathematics as a story of inevitable progress from materiality to intellection, modeling the pattern of education that Proclus wished his own students to follow (Claessens 2015). Proclus's *Commentary* was first printed in a faulty Greek text in 1533, accompanying the first printed edition (*editio princeps*) of the Greek text of the *Elements*. In 1560, from a far better Greek text, the Venetian humanist and mathematician Francesco Barozzi (1537–1604) published a masterful Latin translation, which would be read widely as a source for both the history and philosophy of mathematics (cf. Chapter 4). Apart from providing extensive and important historical information in his commentary, by placing his historical account in the preliminary material before the more technical commentary on Euclid's text Proclus established a pattern for introducing mathematics in a philosophical or humanistic context: to draw students into the subject through a narrative of its development.

Historical narrative and vivid anecdotes could thereby recommend the mathematical sciences to a varied audience. The humanist could be engaged by emphasizing their great antiquity and close connection to liberal pursuits; the theologian by their divine origins and cultivation among the Biblical patriarchs (Adam, Seth, and Abraham); the aspiring courtier or politician could be assured of the value that ancient kings put in the discoveries of Euclid and Archimedes; and the philosopher could learn about the historical connection between mathematics and the schools of Plato and Aristotle.

The astronomer and pioneer in mathematical printing Johannes Regiomontanus began a series of mathematical lectures in Padua in 1464 with an oration on the history of the mathematical sciences (subsequently printed several times). Regiomontanus' account was drawn from popular medieval sources such as the encyclopedic *Etymologies* of the scholar and Archbishop Isidore of Seville (*c.* 560–636). These he supplemented with some recently recovered Greek historical works, but above all with historical data embedded in mathematical works themselves, such as Ptolemy's *Almagest* (Byrne 2006). Philip Melanchthon (1497–1560), Luther's brother in arms and known as the *Praeceptor Germaniae* (teacher of Germany), wrote a much replicated historical essay that helped establish the mathematical sciences at the new Protestant universities in the 1530s (Melanchthon 1999: 98–104). In that essay and other prefaces, Melanchthon first set out many of the historical anecdotes that would become commonplaces in subsequent literary and artistic representations of the mathematical sciences, using them to argue that the highest and truest aim of pure mathematics was philosophical contemplation. In the late sixteenth century, the Jesuit mathematician and astronomer Christoph Clavius (1558–1612), highly esteemed throughout Europe, pointed to the history of the mathematical sciences to establish a permanent place for them in the standard curriculum of the Jesuit Colleges around the world (the *ratio studiorum*).

The mathematically minded humanist Petrus Ramus (1515–1572) in Paris published an entire, very influential book, *Prooemium mathematicum* (Mathematical Introduction) on the history of the mathematical sciences that represented them quite differently from many of his contemporaries in the 1560s, emphasizing their utility for practical ends. In 1570, the young Henry Savile (1549–1622) in Oxford, excellently trained in Greek and mathematics, began his lectures on Ptolemy with a historical narrative that challenged Ramus on almost every point, depicting mathematics as an essentially *theoretical* discipline—while nevertheless exploiting Ramus's *Prooemium* as his main historical source. Thus, history became an obvious means for introducing the mathematical sciences or specific parts thereof to an audience that was predisposed to be rather indifferent towards this realm of knowledge. At the same time such histories were very flexible in their definitions and delineations of the mathematical sciences.

Such written historical accounts belonged most naturally at the beginning of works on the mathematical sciences or specific branches as astronomy or geometry: as prefatory letters and orations. They helped interpret for the readers, in culturally recognizable terms, the challenging material that occupied the rest of the volume. Also, they enticed them to explore a subject they might otherwise have thought was difficult and irrelevant. Savile's extraordinarily erudite introduction to the mathematical sciences established his reputation for learning at Oxford, and later in his life, he returned to his catalogue of worthy mathematicians, in another genre, sprinkling portraits of these figures among the exemplars of learning, in a frieze that decorated (and still decorates) the Reading Room of the new Bodleian Library, for which Savile provided the program (Bullard 1994). Euclid, Ptolemy, Copernicus, and Tycho Brahe appeared in the company of great philosophers, poets, and theologians; and the mathematical sciences took their deserved place among the liberal arts with a clear tendency to supersede them as we shall see in what follows.

THE CULTURAL AND EPISTEMOLOGICAL STATUS OF MATHEMATICS

The prefaces and engraved title pages of books were, of course, tied to the rise of printing. The very appearance of printed mathematical texts—diagrams, clearly delineated theorems and demonstrations, commentary, and lemmas, laid out attractively on the page—carried with it a kind of intellectual authority: here was something that, even if beyond our understanding, was undoubtedly *true*. Authors with modest mathematical aptitude, such as Giordano Bruno, Joseph Scaliger, and Thomas Hobbes, dabbled in mathematics to build up their intellectual prestige—and, not surprisingly, they were castigated for this by more astute mathematicians.

Mathematical knowledge (as found in, say, Euclid's *Elements*) was generally supposed not only to be indubitably true, but to possess the highest degree of certainty humanly attainable. Yet, that was itself a point of contention at times in this period, following the attack of the astronomer and philosopher Alessandro Piccolomini (1508–79) on the certainty of mathematics (1547; cf. Chapter 4). His arguments were attacked, in turn, by several mathematicians, including Francesco Barozzi (in 1560), the Latin translator of Proclus's *Commentary* on Euclid. Such arguments had very little effect on the practice or content of the mathematical sciences; rather, they concerned their reputation, particularly with respect to dialectic and philosophy (cf. Chapter 4).

What, then, were the mathematical sciences *for*? The answer to this question ran the gamut from high Platonic speculation (as, for instance, in Bartolomeo Zamberti's preface to his new humanistic translation of Euclid (1505)); to practical mathematics within a university context (as in Ramus's 1567 *Prooemium mathematicum*); to mathematical knowledge and methods used in the world, for purposes both practical and occult (as in John Dee's *Mathematicall Praeface* to the first English translation of the *Elements* in 1570); to the everyday uses of tradesmen and sailors (as in the vernacular textbooks published across Europe—often originating from the pens of university trained mathematicians). In almost all of these cases, the authors not only were aware of other claims on the nature of the mathematical sciences; they criticized competing representations of other fields of knowledge, and held their own up as the only genuine and worthwhile position (often drawing on the tropes of antiquity or utility explored in this chapter).

In the first prologue to his *Commentary* on Euclid, Proclus emphasized the utility of mathematical knowledge to all of the arts and fields of knowledge. We might expect him to explain how mathematical methods could be used in practical arts that rely on counting, measurement, and weighing, and he does—but only as a brief afterthought to a much more detailed account of the role of mathematics in theology, physics, political philosophy, rhetoric, and poetry. These more unlikely applications of mathematical knowledge became a commonplace in sixteenth century defenses of the mathematical sciences against the charge of their apparent uselessness. In his 1570 lectures, Savile elaborated at length on Proclus's list, similarly insisting that mathematical knowledge was at its most useful when it drew the mind away from matter and into the realm of theoretical contemplation. Ramus, on the other hand, was less impressed by Proclus's rather ethereal notion of "application." He urged his students to take themselves to the heart of Paris to observe merchants and goldsmiths at work. There, they would see the mathematical sciences not only properly applied, but would also see them in what he thought was their most authentic form. He had, in fact, been advertising the utility of the mathematical sciences for several years (Goulding 2018: 358–9). By the time he wrote his *Prooemium* in 1567, he had set this project aside. Nevertheless, he made a strong case for the applicability and utility of the mathematical sciences in all the arts and in ordinary life, which he illustrated through a catalogue of the great mathematical practitioners of his own day. And he urged the princes of Europe—including Queen Elizabeth in England—to establish chairs of mathematics at their nation's universities as a spur to the study of practical mathematics at all levels of society.

Ramus had earlier discussed the need for a chair of mathematics at Oxford and Cambridge with his correspondent John Dee (1527–1608/09) (Ramus 1599: 174–5). Three years after Ramus published his *Prooemium*, Dee himself wrote one of the most important introductions to the mathematical sciences and their utility, in the "Mathematicall Praeface," which appeared at the start of the first published English translation of Euclid (1570) by Henry Billingsley (Dee 1570; Simpkins 1966: 230–3). Dee's long outline of the mathematical sciences was, in effect, yet another elaboration of Proclus's prefatory lectures on geometry. Dee built upon Proclus's account of the utility of mathematics as well as his division of the mathematical sciences first into four parts (the familiar mathematical *quadrivium*), and then into two practical and six applied branches. Dee's own classification added further subdivisions to Proclus's, onto which Dee imposed curiously complicated names ("trochilike," for the study of circular mechanical motions, for example, or "hypsometrie" for that branch of surveying concerned with heights). Like Proclus, he, too, understood the mathematical sciences to occupy a middle place in our mental world, as a facet of the active imagination poised between sense perception and pure intellection, but capable of reaching up and down into both. More than Proclus, however, Dee was genuinely interested in the practical applications of the mathematical sciences. His final mathematical art, "archemastrie," at a first glance seems to be a kind of universal experimental method built on mathematical principles. But in fact it was closely associated in his mind with the magical activities that had begun to occupy him more than his mathematical work (Clulee 1988: 145–76). Thus, even in a work that went much further than many learned introductions to the mathematical sciences in embracing their practical uses, the ultimate goal of "application" was still a way of leaving behind the world, rather than interpreting it with arithmetic and geometry. Dee supplemented his account with a huge fold-out diagram delineating the divisions of the mathematical sciences, his "Groundplat," illustrating on one sheet the reach of the mathematical sciences, from their myriad practical applications to its insight into the divine (Rampling 2011: 140–2).

In seeking patronage, mathematicians inevitably stressed the utility of the mathematical sciences for practical purposes, and almost as inevitably exaggerated their actual accomplishments. The field of optics, a branch of the mathematical sciences under rapid development in this period (Smith 2014), was an especially fertile topic. Even in the thirteenth century, the Franciscan Friar Roger Bacon (*c.* 1219/29–*c.* 1292) had made extraordinary claims about the value of optics in his correspondence with the pope, promoting the use of non-planar mirrors both to see things at a distance and to set distant objects ablaze. Bacon's claims were but one example in a long tradition of imagining the extraordinary capacities of optical objects (Reeves 2008). In the preface to his edition of his father's *Pantometria* in 1571, John Dee's friend and protégé Thomas Digges (*c.* 1546–95) recalled both the supposed feat of Archimedes, in burning the fleet at Syracuse with a parabolic mirror, and his father's own experiments with mirrors. According to Digges, his father was able to use "proportionall Glasses" to read even the inscriptions on coins in distant fields, and to reveal what had been done in "private places" seven miles away; while he had been able to discharge a cannon with the sun's rays from the distance of half a mile. Such feats were of obvious interest to the state, and it is hardly a coincidence that Digges dedicated the volume to Nicholas Bacon, Lord Keeper of the Great Seal. Digges's claims, probably greatly exaggerated, have led to a lively modern debate on the existence of an "Elizabethan telescope." Yet, other mathematical practitioners took notice at the time, both of Digges's claims and their political utility. Around 1580, William Bourne wrote a short treatise to Lord Burghley, rehearsing the optical properties of concave mirrors and convex lenses, hoping that he could be of service to the crown just as he had in a recent work on buoyancy. He speculated that the phenomena recalled by Thomas Digges could be reproduced with a convex lens throwing its image into a concave mirror—which could then throw the same image into a series of other mirrors, until the required size was achieved. But, he confessed, he did not have the funds that Dee and Digges had at hand, so he requested that Burghley patronize his research—an entreaty that seems to have gone unheard (Dupré 2010).

Similar claims, among much else, were made in the most famous book of mathematical marvels, Giambattista della Porta's *Magia naturalis*. This bestseller was published first in a four-book version in 1560, and in a greatly expanded twenty-book version in 1589. The later version was encyclopedic in scope, presenting "experiments" in everything from plant husbandry, to fireworks, to optics. This last topic was the subject of the seventeenth book, which greatly expanded on the fourth book of the 1560 text, and changed its focus: the older text had concerned for the most part well-known mirror illusions, while the newer version extolled the marvels of refraction. In one of the most famous passages, Della Porta explained how to improve the image in a *camera obscura* by means of a lens placed in its aperture—a passage that would later inspire

Kepler in his account of the functioning of the eye (cf. Chapter 5). Mathematical marvels served not only to popularize the mathematical sciences, but by extending and imagining the application of the mathematical sciences, they could stimulate genuine discovery (Borrelli et al. 2017).

THE APPEARANCE OF TEXTBOOKS

The first printed version of Euclid's *Elements*, which appeared in Venice in 1482, published by Erhard Ratdolt (1447–1528), was also one of the most beautiful early editions. Ratdolt published the popular Latin version of the *Elements* originally arranged by Campanus of Novara (*c.* 1220–96), with large margins containing finely cut diagrams. Two subsequent Venetian editions—Joannes Tacuinus's edition of Bartolomeo Zamberti's translation from the Greek (Venice 1505) and Paganinus' edition of text prepared by Luca Pacioli (Venice 1509)—maintained the same very high standards (Thomas-Stanford 1924). Another typographical milestone was the 1570 English translation by Henry Billingsley (d. 1606), which provided fold-out three-dimensional models for all the theorems on solid geometry. These were expensive editions, usually out of the reach of those interested in mathematics. Large editions continued to be produced, but often "added value" for the reader, for example, multiple versions of Euclid's text (such as Jacques Lefèvre d'Étaples 1516 edition, which supplemented Zamberti's humanistically faithful version with the much more familiar and popular Campanus version). Some offered extensive commentaries, such as the editions of the Italian scholars Francesco Maurolico (1494–1575) and Federico Commandino (1506–75) (Rose 1975).

In the opposite direction, editors produced stripped down versions of the *Elements*, containing only the statements of the propositions without their proofs or diagrams. They were encouraged to do so by the erroneous belief that the proofs and diagrams in the received text of Euclid were "commentaries" written by Euclid's fourth-century editor Theon of Alexandria. Ironically this misconception was put into circulation by Zamberti in the preface to his more accurate version of the *Elements* (Goulding 2010). In his bare-bones version of 1549, Ramus tried to make the austerity of his version into a virtue. The absence of demonstrations or diagrams made Euclid affordable to the average university student, and he would learn more by watching his professor draw the diagrams and work through the proofs in front of him than by reading about them in a book. In a classroom in the Jesuit College in Coimbra, diagrams illustrating Euclid's arguments were constantly on view on glazed tiles (azulejos), copied from the Jesuit mathematician André Tacquet's edition of and commentary on Euclid's *Elements* of 1654 (Leitão and Gessner 2014).

The textbook tradition in the mathematical sciences also developed rapidly over this period (cf. Chapter 6). Early printed mathematical books tended to be editions of ancient or medieval texts. Jacques Lefèvre d'Étaples (*c.* 1455–*c.* 1536), for instance, founded his attempted renewal of mathematics in Paris on the revival of the Boethian tradition of arithmetic and geometry, preparing new editions of these staples of the medieval curriculum (Oosterhoff 2018). Oronce Fine (1494–1555), on the other hand, took a different approach from his perch at Paris' *Collège Royal*, publishing textbooks on practical mathematics that were often derided for their over-simplification, but which nonetheless proved enormously popular in both France and England (Marr 2009). In England, Fine's work was eventually eclipsed by the series of textbooks written in English by the Welsh mathematician Robert Recorde, introducing readers to arithmetic, geometry, astronomy, and algebra (Williams 2011). These textbooks were accessibly written as a dialogue between a master and a student and were laid out attractively with clear diagrams. Moreover, they were mathematically innovative and sophisticated. Recorde invented and introduced, for instance, the equals sign, and was thereby able to set out and solve algebraic equations in a much more intuitive form, resembling modern methods (Hughes 1993). Ramus, too, seemed to learn from the developing field of textbook production. His early textbooks on arithmetic and geometry consisted of dense thickets of text, with little recourse (despite Ramus's reputation) to practical mathematics or the student's experience of the world. The final version of his mathematical textbooks, which he published in 1569, illustrated their contents abundantly with practical examples, and pictures of students engaged in surveying and measurement (Goulding 2018). These final versions of Ramist mathematics proved to be enormously popular. Translated into the vernacular, they even formed the basis of elementary mathematical instruction at many Protestant universities.

Even as popular approaches to mathematical education proliferated, the magisterial translations of Federico Commandino and Christopher Clavius still embodied a kind of authority. These large folio volumes, with beautiful diagrams, long-learned commentaries on propositions, and citations from their Greek sources scattered throughout, conveyed a kind of solidity even to those who were barely able to understand their contents. The persuasive rhetoric of presentation led to a crisis in 1594, when the renowned humanist historian and chronographer Joseph Justus Scaliger (1540–1609) published a work entitled *Cyclometrica elementa duo*. This work was lavishly printed in red and black type, with beautifully cut diagrams worked into the text, and propositions stated first in Greek, then Latin, followed by extensive commentary (see Figure 7.2).

Scaliger was not printing an ancient text, however, but his own discovery of the quadrature of the circle. He openly disavowed, in fact, the mathematics of the ancients, attacking Archimedes at length in his preface to the book, rejecting the Greek mathematician's estimate of the ratio of the diameter of the circle to its circumference (between $3\frac{1}{7}$ and $3\frac{10}{71}$). Scaliger believed he had found

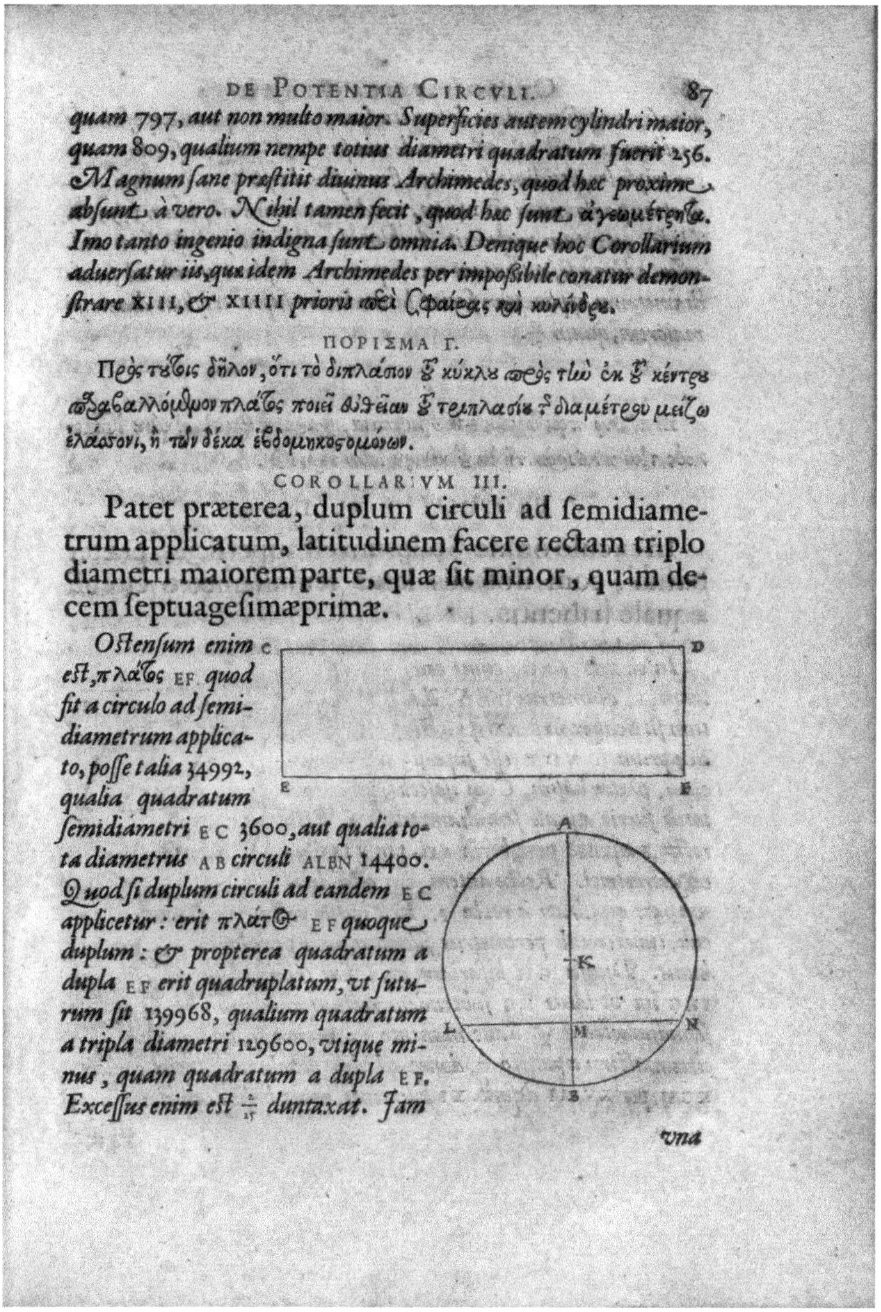

DE POTENTIA CIRCVLI. 87

quam 797, aut non multo maior. Superficies autem cylindri maior, quam 809, qualium nempe totius diametri quadratum fuerit 256. Magnum sane præstitit diuinus Archimedes, quod hæc proxime absunt à vero. Nihil tamen fecit, quod hæc sunt ἀγεωμέτρητα. Imo tanto ingenio indigna sunt omnia. Denique hoc Corollarium aduersatur iis, quæ idem Archimedes per impossibile conatur demonstrare XIII, & XIIII *prioris* περὶ σφαίρας καὶ κυλίνδρου.

ΠΟΡΙΣΜΑ Γ.

Πρὸς τούτοις δῆλον, ὅτι τὸ διπλάσιον τοῦ κύκλου πρὸς τὴν ἐκ τοῦ κέντρου παραβαλλόμενον πλάτος ποιεῖ εὐθεῖαν τοῦ τριπλασίου τῆς διαμέτρου μείζω ἐλάσσονι, ἢ τῶν δέκα ἑβδομηκοστομόνων.

COROLLARIVM III.

Patet præterea, duplum circuli ad semidiametrum applicatum, latitudinem facere rectam triplo diametri maiorem parte, quæ sit minor, quam decem septuagesimæprimæ.

Ostensum enim est, πλάτος EF *quod sit a circulo ad semidiametrum applicato, posse talia 34992, qualia quadratum semidiametri* EC *3600, aut qualia tota diametrus* AB *circuli* ALBN *14400. Quod si duplum circuli ad eandem* EC *applicetur: erit* πλάτος EF *quoque duplum: & propterea quadratum a dupla* EF *erit quadruplatum, vt futurum sit 139968, qualium quadratum a tripla diametri 129600, vtique minus, quam quadratum a dupla* EF. *Excessus enim est* $\frac{1}{15}$ *duntaxat. Iam*

vna

FIGURE 7.2. Scaliger: *Cyclometrica elementa* duo, p. 87. Max Planck Institute for the History of Science, Berlin.

a direct and exact quadrature of the circle, which amounted to asserting that $\pi = \sqrt{10}$. Mathematicians throughout Europe published refutations of Scaliger's work, fearing that the public would be swayed not only by Scaliger's formidable reputation as a scholar, but by the authoritative appearance of the work. The Flemish mathematician Adriaan van Roomen (1561–1615) went so far as to suggest the formation of a panel of recognized mathematicians who would approve or suppress every book that purported to contribute to mathematics. The *Cyclometrica* controversy, which provoked widespread debate about the role of humanism in the mathematical sciences, the value of representation, the skills proper to a mathematician, and the authority earned thereby, helped to create in Europe the category of a professional mathematician. The learned world very slowly began to define "mathematics" as a specialized field requiring training that was beyond the generally educated classical scholar (Goulding 2005).

VISUAL REPRESENTATIONS

All these developments unfolded in a world of intellectual and social elites who were highly sensitive to imagery, be it religious or secular. Indeed, representations of arithmetic, astronomy, geometry, and music, the four mathematical sciences of the quadrivium, had been widely spread throughout Europe since the Middle Ages as part of rather popular and ubiquitous depictions of the seven liberal arts (the other three, the trivium, being, dialectic, grammar, and rhetoric). The seven liberal arts had become part of university education due to their propaedeutic value for the study of law, medicine, or theology, the three backbones of medieval and early modern universities. Knowledge of them and their representations was natural for the educated. Thus, it is not surprising that a cycle of the seven liberal arts adorned the tomb of pope Sixtus IV (1414–84) in St. Peter in Rome, indicating that he had been a generous patron of knowledge. During the sixteenth and seventeenth centuries printed cycles of these seven branches of learning spread all over Europe. Usually, the personifications of arithmetic, astronomy, geometry, and music could easily be decoded because they carried standard attributes and were often also associated with well-known scholars from antiquity. Arithmetic, associated with Pythagoras or Boethius, would count coins or hold a table with numbers. Astronomy/astrology, associated with Ptolemy, would often be depicted with wings, a sphere, an astrolabe, or measuring instruments, and would be looking upwards. Geometry, associated with Euclid, would be looking downwards, holding a circle, an angle, a measuring rod, or a table with geometric drawings. Finally, music, associated with Jubal or Pythagoras, would play a musical instrument. These standard attributes can be seen in the engravings of arithmetic and geometry in Lukas Kilian's liberal arts cycle published in 1606 or the arithmetic Cornelis

Cort engraved in 1565 (after Frans Floris), an accounting scene with references to Abraham and Pythagoras as distinguished founders of arithmetic (see the books in front of the table; Figure 7.3). In 1630, when the Ulm mathematician and fortress architect, Johannes Faulhaber (1580–1635), published his *Ingenieurs-Schul* (School for Engineers) he presented eight additional personifications of mathematical sciences on his frontispiece (see Figure 7.4), reflecting the beginning and from then ever-increasing differentiation within the mathematical sciences that finally led to the establishment of specific scientific disciplines until the nineteenth century.

In the first row Geometry, Topography (usually a subdiscipline of geography), Geography itself and Arithmetic can be seen; in the second row there are Music, Gnomonics (the theory of sundials), Trigonometry and Optics; and at the back we find Architecture, Astronomy, Logarithms, and Mechanics. This arrangement is a representation of the theoretical and practical scope of the mathematical sciences.

A century before Faulhaber one of the most striking illustrations of the new role the mathematical sciences aspired to, in the context of the traditional

FIGURE 7.3. Three men read while a women writes numerals; representing arithmetic. Engraving by C. Cort, 1565, after F. Floris, *c.* 1557. Wellcome Collection. Public Domain Mark.

FIGURE 7.4. Faulhaber, Johannes. 1630. *Ingenieurs-Schul / Erster Theyl*. Herzog August Bibliothek Wolfenbüttel: A: 26.1 Bell. (1).

FIGURE 7.5. Tartaglia, Niccolò: *Nova scientia*. Vinegia: S. da Sabio. Courtesy of the Smithsonian Libraries.

hierarchy of knowledge, is found as the frontispiece to the *Nova Scientia* (New Science, Venice 1537) by the self-taught engineer and mathematician Niccolò Tartaglia (1499/1500–77). In this famous woodcut (Figure 7.5), Euclid is shown as the gatekeeper to the walled garden of knowledge, allowing only those to enter whom he admits (a conceit underlined by the figure unsuccessfully trying to break into the garden on a ladder).

Within the garden stands a crowd of mathematical arts, all individually labelled, with Geometry and Arithmetic in the foreground. The crowd of personified arts are raptly watching the demonstration of a cannon, the trajectory of which has been drawn using Tartaglia's mathematical principles. But the foremost figure, standing between Arithmetic and Geometry, is not Mathematics, but Tartaglia himself. Behind them stand Aristotle and Plato, the latter adorned with the famous tag, "Let no one enter here who is ignorant of geometry," guarding the gate into a smaller enclosure, where Philosophia sits enthroned. The image emphasizes a movement towards knowledge: a pair of seekers are shown outside the walls, then are ushered in by Euclid, and finally move beyond Tartaglia and the mathematical arts—and the demonstration of the cannon—towards the sanctuary of philosophy. The inscription at the bottom made this explicit: "The mathematical sciences say, whoever strives to investigate the reasons to think, study us, this way is open to everyone." Tartaglia's image not only represents the traditional role of Euclid and the mathematical arts in the journey to philosophy, but places Tartaglia himself and his "new science" at the center of a vividly depicted pedagogical drama (Cuomo 1997).

In the fine arts *Mathematica* was often represented bearing wings on her head. Such a portrait appeared in the *Iconologia* of Cesare Ripa (?–1622), an influential iconological treatise serving as vademecum for many artists throughout the early modern period. In the *Iconologia*'s first illustrated edition of 1603, *Mathematica* is accompanied by a compass, a table with numbers and figures, a sphere and a pyramid (Figure 7.6). But Ripa also suggested that the wings enabled *Mathematica* to fly to the "contemplation of abstract things." The personification of *Mathematica* as codified by Cesare Ripa in the sixteenth century had spread all over Europe by the seventeenth century.

Mathematicians from antiquity—mostly Greek antiquity, but also from Biblical Times—such as Abraham, Archimedes and Euclid were often referred to visually. Archimedes became particularly prominent because he could be used to symbolize both pure and mixed mathematics. We often find them together in engraved title pages such as the one of Samuel Marolois's *Opera mathematica* (The Hague 1614), often with a special emphasis on Archimedes in the service of the military. Archimedes offers an excellent example of various inflated claims often made for the mathematical sciences. The dramatic stories of Archimedes destroying Roman ships with the help of a burning mirror or lifting them out of the water with giant machines at the siege of Syracuse (213/212 BCE) were repeated endlessly, sometimes with the explicit promise that the patronage of the mathematical sciences would lead to similar and even better instruments of war. In this process, Archimedes was fashioned into both an extraordinarily gifted pure mathematician who had given an excellent approximation of π and a highly creative engineer whose ingenious inventions were of immense value in war and peace. While Archimedes's usefulness to the military was an important aspect and easy to depict, the mathematical achievements of Archimedes and Euclid, and their undiminished cultural importance, were more difficult to present visually. However, they were both highly suitable for emphasizing the antiquity of the mathematical sciences, as can be seen in a series of frontispieces used by Oxford University Press for editions of Euclid in 1703, Apollonius in 1710, and Archimedes in 1792 (Figure 7.7). The

308 ICONOLOGIA.

MATEMATICA.

nifeſtarla in qualche modo, e di quì ſono nate l'inuentioni di muſiche, di proſpettiua, di Architettura, di Geometria, d'Aritmetica, e d'altre profeſſioni, che tutti date alle Stampe, & cauate da' principij di queſta ſcienza continuamente recono guſto alli ſtudioſi con ſodisfattione de gl'autori, i quali per queſti mezi, come per ampia ſcala ſagliono alla fama, & all'immortalità.

Tali habbiamo molti de gl'antichi, & non pochi che viuono à gloria dell'età noſtra, frà quali hanno luogo Chriſtoforo Clauio, Giouan Paolo Vernalione, Giouan Battiſta Raimondo, Luca Valerio, Federico Metio, Pietro Maillardi, Ceſare Ruida, Camillo Agrippa, & molti altri chè con eſquiſita ſcienza, & con fondamento che viuamente poſſiedono, in premio delle fatiche loro in dono in queſta profeſſione al noſtro ſecolo fama

ſmarrita

FIGURE 7.6. Cesare Ripa: *Iconologia*. © Universitätsbibliothek Heidelberg.

FIGURE 7.7. *Euclidis quae supersunt omnia*, David Gregory (ed.). Courtesy of The Linda Hall Library of Science, Engineering & Technology.

only variations between the different versions were the geometric drawings on the left of the three men on the beach. A quotation from Vitruvius below the picture explained the scene: "Aristippus the Socratic philosopher, when shipwrecked on the island of Rhodes, noticed geometrical diagrams drawn on the beach and said to his companions, 'We can hope for the best for I see the signs of men.'" This anecdote had been a staple of prefaces to mathematical works, starting with Melanchthon's much-read exhortation to mathematics. Clearly, mathematics stood for civilization and, even more, cultural superiority.

The superior cultural status of pure mathematics and the mathematical sciences in general as reflected in this image had already moved center stage by the mid-seventeenth century. In the engraved title page of his textbook of the mathematical sciences for use in Jesuit colleges, the *Cursus mathematicus* of 1661, the Jesuit Gaspar Schott (1608–66) had a crowned personification of the mathematical sciences present a dedication copy of his book to the Holy Roman Emperor, Leopold I. In this way, Schott emphasized that *Mathematica* was the queen among the sciences, worthy to communicate with the Emperor himself. A few years later Schott's French fellow Jesuit, Claude-François Milliet de Chales (1621–78), in the dedication to his own *Cursus mathematicus* of 1674, confidently asserted that "the other disciplines are plebeian; the mathematical sciences are royal" ("Plebeiae sunt ceterae disciplinae, mathesis Regia").

The whole gamut of arguments for the antiquity, nobility and utility of the mathematical sciences not only played out visually on frontispieces and engraved title pages but also in painted portraits of mathematicians. Two famous examples are the portrait of the Italian mathematician Luca Pacioli by Jacopo de' Barbari, *c.* 1495 (cf. the cover of this volume), and Hans Holbein's *Ambassadors*, a portrait of two noblemen from 1533, richly decorated with mathematical instruments. Another genre were portraits of imagined figures from antiquity or mythology. Two famous Greek mathematicians were depicted, for example, Euclid in a painting by Joos van Wassenhove from 1474, and in 1620, Domenico Fetti's portrait of Archimedes. One also finds numerous engraved images (e.g., representations of the seven liberal arts) as well as imagery on mathematical instruments. These typically refer to Greek antiquity (Archimedes, Atlas, etc.) and the Bible (Abraham, etc.), to trade and war, or to royalty engaging in the mathematical sciences. Often the history and power of the mathematical sciences played out visually at courts, as in Raphael's magnificent fresco *School of Athens* (1508–11), which depicts several famous Greek mathematicians in the company of Aristotle and Plato. Pope Julius II commissioned this work for the Vatican. Other examples are the frescos Giulio Parigi painted in the Uffizi in Florence in 1599/1600 showing historical scenes from the mathematical sciences as well as the miracles that Archimedes performed at the siege of Syracuse.

At courts and among the wealthy, mathematical instruments, representations of mathematical objects, and ways to make mathematical knowledge visible (models) became very fashionable in the sixteenth and seventeenth centuries. Among these, mathematical instruments, or ideas made of brass as they have been called, such as astrolabes, surveying instruments, or astronomical clocks turned into collectibles. Often they would be adorned with imagery, some pointing to possible applications pertaining to the mathematical sciences, others with designs deeply rooted in history, constructing specific lines of tradition, and legends, both biblical and Greek (Ellinghaus and Remmert 2022). In Germany, Landgrave William IV of Hessen-Kassel (1532–92), known as the Ptolemy of Kassel because of his astronomical expertise, asked his mathematician and clockmaker Jost Bürgi (1552–1632), a multitalented Swiss craftsman who also happened to have been the co-inventor of logarithms, to build an astronomical clock. The clock was finished in 1591 and is rather small, based on a square of 15.4 cm and barely 11 cm high. The imagery on the four sides shows two episodes from the history of geometry and astronomy on each side, sketching out a line of tradition from ancient times to the sixteenth century (Gaulke 2015). The four panels depict, respectively, three patriarchs (Abraham, Adam, and Seth) and Thales; Euclid and Archimedes; Hipparchus and Ptolemy; finally the medieval patron of astronomy, King Alfonso of Castile, and Copernicus. It is obvious that Hipparchus and Ptolemy; King Alfonso and Copernicus stand for a tradition in astronomy, whereas Thales, Euclid, and Archimedes stand for a tradition in geometry, and that both traditions reach back to Greek antiquity. It is, however, interesting to note that both lines have their origin in biblical times with the three patriarchs, a reference to legends that Abraham had instructed the Egyptians in arithmetic and astronomy while Adam and his son Seth had inscribed the antediluvian knowledge on the two pillars to safeguard it from catastrophes to come. During the Middle Ages specific fields of knowledge had been ascribed to the two pillars: geometry and astronomy, often depicted in manuscripts and paintings (cf. Figure 7.1). Thus, the essential point of the pictorial program on the clock beginning with the patriarchs is that the origin of astronomy and geometry in ancient times is twofold: it derived from both Greek antiquity and the patriarchal world of the Old Testament.

THE UTILITY OF THE MATHEMATICAL SCIENCES

During the fifteenth century, proponents of the mathematical sciences typically singled out three aspects as praiseworthy: their antiquity, their propaedeutic value for the study of philosophy, and their practical advantage for the community (Swerdlow 1993). During the sixteenth century the strategies used to legitimize the mathematical sciences became fairly standardized, and drew on a common reservoir of arguments (the practical and propaedeutic role of the mathematical sciences) and examples (Archimedes's burning mirrors being among the most popular). The educational value of the mathematical sciences, to which their epistemological status was closely related, was usually seen to be in their importance for training the mind and in their recreational potential, but not often in their necessity or worth for other fields of knowledge, such as philosophy, medicine, law, or theology. This situation gradually changed until the first half of the seventeenth century, however, when mathematicians began to argue that the mathematical sciences deserved a new position in the traditional hierarchy of knowledge, namely at the very top. They based this claim on a twofold argument, first by emphasizing the absolute certainty of mathematical knowledge—the topic that had been so hotly debated in the *quaestio de certitudine mathematicarum* (cf. Chapter 4)—and second, on the utility of the mathematical sciences (cf. Chapter 6).

While it was rather difficult to argue visually for the certainty of pure mathematics, the claim that the mathematical sciences were applicable could easily be illustrated almost everywhere. Martinus Hortensius (Maarten van den Hove, 1605–39), in his Amsterdam inaugural lecture of 1634 on the *Dignity and the Usefulness of the Mathematical Sciences*, gave an exhaustive overview of the relevant arguments (Imhausen and Remmert 2006). Hortensius's speech is indicative of the fact that by the mid-seventeenth century the mathematical sciences and trends of mathematization had become highly fashionable and respectable in Europe. Drawing on various predecessors such as Ramus, Hortensius gave a long list of examples of the utility of the mathematical sciences. He first discussed how the four university faculties—philosophy, theology, law, and medicine—all depended on the mathematical sciences. However, Hortensius also listed many branches of knowledge depending on mathematical expertise, such as practical arithmetic, geodesy, military architecture, mechanics and statics, music, optics, astronomy, geography and navigation. In the context of Amsterdam's reputation as a leading center of trade, Hortensius paid particular attention to the advantages of practical arithmetic, geography, and navigation. Putting his message about the indispensability of the mathematical sciences in a nutshell he said: "so much advantage have the mathematical sciences contributed to navigation, navigation to trade, and trade to the solid and firm prosperity of our country" (Imhausen and Remmert 2006: 129). Just in this vein, most of the visual strategies to illustrate the utility of the mathematical sciences pertained to war and trade: they were essential to the military crafts in war and they substantially contributed to the growth and safeguarding of prosperity in times of peace (cf. Chapter 6). The theme of warfare was often incorporated in title engravings of mathematical publications, by means of fortification or ballistic scenes for instance. A typical example is the frontispiece of the *Mathematical Dictionary* (Dictionaire mathematique, Amsterdam and Paris 1691; see Figure 7.8) by the French mathematician and prolific author, Jacques Ozanam (1640–1718).

FIGURE 7.8. Jacques Ozanam: *Dictionnaire mathématique*. Herzog August Bibliothek Wolfenbüttel: A: 7.2 Geom.

He conceived this book as a small encyclopedia of the mathematical sciences consisting of individual overviews of specific branches, including algebra, geometry, optics and military architecture. In fact, the military context was particularly striking. In the preface, Ozanam, besides praising the general advantages of the mathematical sciences, quite clearly promoted their military relevance, as reflected in the frontispiece. The scene at first seems to represent the whole extent of the mathematical sciences, but the dominant figure is the personification of geometry, sitting enthroned in the right of the picture. She is explaining a fortress plan to a female figure, who presumably represents military architecture or possibly a representation of Athena, the goddess of war. The special association between the personification of geometry and the art of fortification can also be seen in the battlement-like shapes of her crown, stylizing geometry as an absolutely essential queen of the art of warfare. In the background, through the arches, artillery can be seen, adding to the war-oriented character of the picture.

On the other hand, the validity of the mathematical sciences for trade and prosperity was well known and had a special significance, in particular in the trading nations of England and the Dutch Republic. Thus, strategies for the legitimization of the mathematical sciences in terms of trade and prosperity had excellent chances of succeeding within the context of the commercial competition between the Dutch and the English. The frontispiece of the *New Systeme of the Mathematicks* of Jonas Moore (1617–79) published in 1681, illustrates this nicely (Figure 7.9).

This work covered, amongst other things, arithmetic, practical geometry, trigonometry, cosmography, and navigation. The frontispiece deals with the close link between these fields and trade, thereby stressing the utility of the mathematical sciences in the world at large and, of course, that of the book itself. The framed vignette shows a group of men gathered around a globe. The numerous mathematical instruments and drawings they hold in their hands, or which lay scattered around, indicate that they are engaged with the mathematical sciences. In the back of the room, the sea and commerce can be seen through a window. On the right and the left of the window frame, Neptune, Lord of the seas, and Aeolus, the keeper or god of the winds, watch over the assembly. As an unobtrusive, but expressive reference to the rewards of maritime trade, Aeolus is shown smoking a pipe. Thus, the basis of the profitable business of tobacco trade, so the frontispiece suggests, was the mathematical sciences. Equating the mathematical sciences with trade and prosperity was a common motif, not only among authors but also among publishers in advertising their products.

Finally, a special form of utility lay in the potential the mathematical sciences had to entertain. One finds this recreational aspect quite often in texts, but also the interest in modifying the natural world into a more pleasurable place. The great progenitor of the latter theme was Giambattista Della Porta's *Magia*

FIGURE 7.9. Jonas Moore: *A New Systeme of the Mathematicks*, London 1681. ETH-Bibliothek Zürich: Rar 5492.

naturalis, already mentioned above. The genre continued to be popular in the seventeenth century, starting with Claude-Gaspar Bachet de Méziriac's *Problèmes plaisans et délectables* (Pleasant and Delectable Problems, Lyon 1612), and the tradition of the *Récréations mathématiques* (Mathematical Recreations), first published in 1624 and then expanded by a host of notable editors and mathematicians, including Descartes's optical collaborator Claude Mydorge (Heeffer 2004). Such booklets included mathematical problems as well as experiments—often from optics, following the pattern set by Della Porta—and quickly spread over Europe circulating the message of the importance and utility of the mathematical sciences. In the natural world, the power of the mathematical sciences to entertain could be seen in early modern gardens, where they often provided a garden with much entertainment and delight. Music and acoustical effects, fountains and sundials, geometrical forms, automata, and many other things derived from the mathematical sciences and turned the world into a more pleasurable place. This can be seen in the frontispiece of the *Aerarium Philosophiae Mathematicae* (Treasure Chest of the Mathematical Sciences) by the Jesuit Mario Bettini (1582–1657). This presented a garden full of entertainments made possible through the mathematical sciences: forms, symmetries, fountains, architecture, and perspectives become a setting for demonstrations of optical illusions, musical performances, and the display of mathematical devices and diagrams (Remmert 2016).

Utility and applicability are, of course, essential to the success story of the mathematical sciences since the seventeenth century. Francis Bacon with his emphasis on scientization ("scientia est potentia") and Galileo Galilei with his emphasis on mathematization (the book of nature "is written in the language of mathematics"), offered a fundamental promise with respect to the positive future effects of science and mathematics. This new role of the mathematical sciences, so crucial in the emergence of the "new science" of Galileo, Kepler and Newton, was often proudly presented visually and textually. On the title page of his *Il Saggiatore* (The Assayer) of 1623, Galileo, who was very much aware of the value of visual communication in his publications, expertly illustrated this development (Figure 7.10).

The engraved title page consists of an architectural arrangement, divided in a classical manner into three horizontal sections. In the middle section, the panel containing the title and the name of the author was flanked by personifications of Natural Philosophy and Mathematica, the latter symbolizing the mathematical sciences. Relations between these two fields at the beginning of the seventeenth century had not been free from tensions (cf. Chapter 4); in particular, the traditional precept of many Aristotelian scholars that mathematical methods had no application in natural philosophy was increasingly questioned. Galileo took up this disagreement in the engraved title page. Natural Philosophy and Mathematica stand equally on each side of the middle

FIGURE 7.10. Galileo Galilei: *Il Saggiatore*, Rome 1623. Photo by SSPL / Getty Images.

panel. Natural Philosophy here represents the dusty outdated book-knowledge associated with the faulty worldview, and Mathematica faces her, crowned and holding a pair of open dividers pointing upwards. This gesture unmistakably symbolizes her claim to theoretical knowledge beyond its mere usefulness for taking measurements of the heavens, the only type of knowledge natural philosophers had been willing to concede to her in the past.

CONCLUSION

In this period, the field of the mathematical sciences was changing rapidly throughout Europe, developing far beyond the traditional contents of the medieval quadrivium. Once Greek and Arabic texts were rediscovered, published, and translated, early modern mathematicians developed their own insights and made original discoveries far beyond the achievements of their predecessors. Yet, the then still emerging mathematical sciences had not yet found an obvious cultural role. The representations of the mathematical sciences, whether in verbal descriptions or visual depictions, helped to explore and advertise their possibilities, often in terms that seem very odd to us—and often (ironically, given the technical advances in mathematical knowledge) in ways that emphasized their antiquity or continuity with non-mathematical culture. At the same time, as the mathematical sciences became ever more the province of the specialist, mathematicians sought ways to depict themselves as essential participants in the mathematical enterprise, whether as conduits of mathematical expertise for the benefit of the public, or as the only credible practitioners of the art.

BIBLIOGRAPHY

Andersen, Kirsti (2006), *The Geometry of an Art: The History of the Mathematical Theory of Perspective from Alberti to Monge*, New York: Springer.
Apollonius of Perga (1566), *Conicorum libri quattuor [. . .] quae omnia nuper Federicus Commandinus Urbinas mendis quamplurimis expurgata è Graeco convertit, & commentariis illustravit*, Bologna: Alessandro Benacci.
Apollonius of Perga (1675), *Apollonii Conica: Methodo nova illustrata, & succinctè demonstrata*, Per Isaacum Barrow, London: William Godbid.
Ariew, Roger, ed. (2000), *René Descartes: Philosophical Essays and Correspondence*, Indianapolis, IN: Hackett.
Arndt, Hans Werner (1971), *Methodo scientifica pertractatum*, Berlin: de Gruyter.
Baron, Margaret ([1969] 2004), *The Origins of the Infinitesimal Calculus*, Oxford: Pergamon Press.
Beeley, Philip (2019), "Practical Mathematicians and Mathematical Practice in Later Seventeenth-Century London," *British Journal for the History of Science*, 52 (2): 225–48.
Bennett, Jim (1986), "The Mechanics' Philosophy and the Mechanical Philosophy," *History of Science*, 24: 1–28.
Bennett, Jim (1991), "The Challenge of Practical Mathematics," in Stephen Pumfrey, Paolo L. Rossi, and Maurice Slawinski eds., *Science, Culture and Popular Belief in Renaissance Europe*, 176–90, Manchester: Manchester University Press.
Bertoloni Meli, Domenico (1993), *Equivalence and Priority: Newton versus Leibniz*, Oxford: Clarendon Press.
Bertoloni Meli, Domenico (2006), *Thinking with Objects: The Transformation of Mechanics in the Seventeenth Century*, Baltimore, MD: Johns Hopkins University Press.
Biagioli, Mario (1989), "The Social Status of Italian Mathematicians, 1450–1600," *History of Science*, 27 (1): 41–95.
Biagioli, Mario (1993), *Galileo, Courtier: The Practice of Science in the Culture of Absolutism*, Chicago, IL: University of Chicago Press.

Biggs, Norman (2019), "John Reynolds of the Mint: A Mathematician in the Service of King and Commonwealth," *Historia Mathematica*, 48: 1–28.

Bodenmann, Siegfried, and Anne-Lise Rey, eds. (2013), "La guerre en lettres: La controverse scientifique dans les correspondances des Lumières," *Revue d'histoire des sciences*, 66: 233–48.

Borelli, Giovanni Alfonso (1989), *On the Movement of Animals*, trans. P. Maquet, Berlin: Springer.

Borrelli, Arianna, Giora Hon, and Yaakov Zik, eds. (2017), *The Optics of Giambattista Della Porta (ca. 1535–1615): A Reassessment*, Cham: Springer.

Bos, Henk J. M. ([1980] 2000), "Newton, Leibniz and the Leibnizian Tradition," in I. Grattan-Guinness ed., *From the Calculus to Set Theory, 1630–1910*, 49–93, London: Duckworth.

Bos, Henk J. M. (2001), *Redefining Geometrical Exactness: Descartes' Transformation of the Early Modern Concept of Construction*, New York: Springer.

Brentjes, Sonja (2018), *Teaching and Learning the Sciences in Islamicate Societies (800–1700)*, Turnhout: Brepols.

Brioist, Pascal (2015), "La guerre et les sciences à la Renaissance," in Stéphane van Damme ed., *Histoire des sciences et des savoirs, vol. 1: De la Renaissance aux Lumières*, 11–131, Paris: Le Seuil.

Bruce, Susan, ed. (2009), *Three Early Modern Utopias: Thomas More: Utopia / Francis Bacon: New Atlantis / Henry Neville: The Isle of Pines*, Oxford: Oxford University Press.

Bullard, M. R. A. (1994), "Talking Heads: The Bodleian Frieze, its Inspiration, Sources, Designer and Significance," *Bodleian Library Record*, 14: 461–500.

Burnett, Charles, and Danielle Jacquart, eds. (2005), *Scientia in margine: Études sur les marginalia dans les manuscrits scientifiques du Moyen Âge à la Renaissance*, Geneva: Droz.

Byrne, James Steven (2006), "A Humanist History of Mathematics? Regiomontanus's Padua Oration in Context," *Journal of the History of Ideas*, 67 (1): 41–61.

Cajori, Florian ([1928/9] 1933), *A History of Mathematical Notations*, New York: Dover.

Cardano, Girolamo (1545), *Artis magnae, sive de regulis algebraicis*, Liber unus, Nuremberg: Johann Petreius.

Centre d'études et de recherches médievales de l'Université de Lyon II, ed. (1978), *La complainte de François Garin, marchand de Lyon (1460). Édition critique*, Lyons: Presses Universitaires de Lyon.

Chalmers, Alan F. (2018), *One Hundred Years of Pressure: Hydrostatics from Stevin to Newton*, New York: Springer.

Charlton, Kenneth (2007), *Education in Renaissance England*, London: Routledge.

Chrisomalis, Stephen ([2009] 2011), "The Cognitive and Cultural Foundations of Numbers", in Eleanor Robson and Jacqueline Stedall eds., *The Oxford Handbook of the History of Mathematics*, 495–518, Oxford: Oxford University Press.

Claessens, Guy (2015), "Reception and the Textuality of History: Ramus and Kepler on Proclus' History and Philosophy of Geometry," in André Lardinois, Sophie Levie, Hans Hoeken, and Christoph Lüthy eds., *Texts, Transmissions, Receptions: Modern Approaches to Narrative*, Radboud Studies in Humanities, #1, 281–94, Leiden: Brill.

Clulee, Nicholas H. (1988), *John Dee's Natural Philosophy: Between Science and Religion*, London: Routledge.

Cohen, H. Floris (1994), *The Scientific Revolution: A Historiographical Inquiry*, Chicago, IL: University of Chicago Press.

Cohen, H. Floris (2010), *How Modern Science Came into the World: Four Civilizations, One 17th Century Breakthrough*, Amsterdam: Amsterdam University Press.

Cohen, I. Bernard (2005), *The Triumph of Numbers*, New York: W.W. Norton.

Copernicus, Nikolaus ([1543] 1992), *On the Revolutions*, trans. and comm. Edward Rosen, Baltimore, MD: Johns Hopkins University Press.

Cormack, Lesley B. (1997), *Charting and Empire: Geography at the English Universities*, Chicago, IL: University of Chicago Press.

Cormack, Lesley B. (2017), "Introduction: Practical Mathematics, Practical Mathematicians, and the Case for Transforming the Study of Nature," in Lesley B. Cormack, Steven A. Walton, and John A. Schuster eds., *Mathematical Practitioners and the Transformation of Natural Knowledge in Early Modern Europe*, Studies in History and Philosophy of Science, #45, 1–10, Dordrecht: Springer.

Cormack, Lesley B., Steven A. Walton, and John A. Schuster, eds. (2017), *Mathematical Practitioners and the Transformation of Natural Knowledge in Early Modern Europe*, Studies in History and Philosophy of Science, #45, Dordrecht: Springer.

Corry, Leo (2015), *A Brief History of Numbers*, Oxford: Oxford University Press.

Cotter, Charles (1978), "Early Tabular, Graphical and Instrumental Methods for Solving Problems of Plane Sailing," *Revista da Universidade de Coimbra*, 26: 3–20.

Cuomo, Serafina (1997), "Shooting by the Book: Notes on Niccolò Tartaglia's *Nova Scientia*," *History of Science*, 35 (2): 155–88.

Daston, Lorraine (1988), *Classical Probability in the Enlightenment*, Princeton, NJ: Princeton University Press.

De Pace, Anna (1993), *Le matematiche e il mondo*, Milan: FrancoAngeli.

Dear, Peter R. (1995), *Discipline & Experience: The Mathematical Way in the Scientific Revolution*, Chicago, IL: University of Chicago Press.

Dear, Peter R. (2001), *Revolutionizing the Sciences: European Knowledge and its Ambitions, 1500–1700*, Basingstoke: Red Globe Press.

Dee, John ([1570] 1975), *The Mathematicall Praeface to the Elements of Geometry of Euclid of Megara*, introduction by A. Debus, New York: History of Science Publications.

Dee, John (1851), *Autobiographical Tracts of Dr. John Dee Warden of the College of Manchester, Chetham Miscellanies*, vol. 1, Manchester: Chetham Society.

Denniss, John ([2009] 2011), "Learning Arithmetic: Textbooks and Their Users in England 1500–1900," in Eleanor Robson and Jacqueline Stedall eds., *The Oxford Handbook of the History of Mathematics*, 448–67, Oxford: Oxford University Press.

Desaguliers, John Theophilus (1744), *A Course of Experimental Philosophy*, vol. 2, London: W. Innys, M. Senex, and T. Longman.

Descartes, René ([1637] 1954), *The Geometry of René Descartes with a Facsimile of the First Edition*, ed. and trans. David Eugene Smith and Marcia Latham, New York: Dover.

Descartes, René (1955), *The Philosophical Works of Descartes*, trans. Elizabeth Sanderson Haldane and G. R. T. Ross, 2 vols, New York: Dover.

Descartes, René (1985–8), *The Philosophical Writings of Descartes*, trans. John Cottingham, Robert Stoothoff, and Dugdale Murdoch, Cambridge: Cambridge University Press.

Descola, Philippe ([2005] 2013), *Beyond Nature and Culture*, trans. J. Lloyd, Chicago, IL: University of Chicago Press.

Drake, Stillman (1957), *Discoveries and Opinions of Galileo*, New York: Doubleday.

Dugas, René ([1955] 1988), *A History of Mechanics*, trans. J. R. Maddox, New York: Dover.

Dupré, Sven (2010), "William Bourne's Invention: Projecting a Telescope and Optical Speculation in Elizabethan England," in Albert Van Helden, Sven Dupré, Rob van Gent, and Huib Zuidervaart eds., *The Origins of the Telescope*, 129–46, Amsterdam: Royal Netherlands Academy of Arts and Sciences Press.

Ellinghaus, Julia, and Volker Remmert (2022), "Visual Worlds on Early Modern Scientific Instruments: Types and Messages," in Ruth Noyes ed., *Reassessing Epistemic Images in the Early Modern World*, 153–74, Amsterdam: Amsterdam University Press.

Epstein, Stephan R., and Maarten Prak, eds. (2008), *Guilds, Innovation, and the European Economy, 1400–1800*, Cambridge: Cambridge University Press.

Favino, Federica (2017), "Gaspare Berti's Legacy: The 'Mathematicalls' in Baroque Rome," *Nuncius: Journal of the Material and Visual History of Science*, 32 (3): 709–73.

Field, J. V. (1997), *The Invention of Infinity: Mathematics and Art in the Renaissance*, Oxford: Oxford University Press.

Fischer, Hubertus, Volker R. Remmert, and Joachim Wolschke-Bulmahn, eds. (2016), *Gardens, Knowledge, and the Sciences in the Early Modern Period*, Trends in the History of Science, Basel: Birkhäuser.

Folkerts, Menso, and Jan Pieter Hogendijk, eds. (1993), *Vestigia Mathematica: Studies in Medieval and Early Modern Mathematics in Honour of H.L.L. Busard*, Amsterdam: Rodopi.

Foucault, Michel ([1966] 1970), *The Order of Things: An Archaeology of the Human Sciences*, London: Tavistock.

Gaida, Margaret (2016), "Reading 'Cosmographia': Peter Apian's Book-Instrument Hybrid and the Rise of the Mathematical Amateur in the Sixteenth Century," *Early Science and Medicine*, 21 (4): 277–302.

Galilei, Galileo ([1638] 1954), *Dialogues Concerning Two New Sciences*, trans. Henry Crew and Alfonso de Salvio, New York: Dover.

Gaulke, Karsten (2015), "Perfect in Every Sense: Scientific Iconography on an Equation Clock by Jost Bürgi and the Self-Understanding of the Astronomers at the Kassel Court in the Late 1580s," *Nuncius: Journal of the Material and Visual History of Science*, 30 (1): 37–74.

Gerbino, Anthony, ed. (2014), *Geometrical Objects: Architecture and the Mathematical Sciences 1400–1800*, Cham: Springer.

Gonzalez de Leon, Fernando (1996), "'Doctors of the Military Discipline': Technical Expertise and the Paradigm of the Spanish Soldier in the Early Modern Period," *Sixteenth Century Journal*, 27 (1): 61–85.

Gouk, Penelope (1988), *The Ivory Sundials of Nuremberg 1500–1700*, Cambridge: Whipple Museum.

Goulding, Robert (2005), "Polemic in the Margin: Henry Savile against Joseph Scaliger's Quadrature of the Circle," in Charles Burnett and Danielle Jacquart eds., *Scientia in margine: Études sur les marginalia dans les manuscrits scientifiques du Moyen Âge à la Renaissance*, 241–60, Geneva: Droz.

Goulding, Robert (2010), *Defending Hypatia: Ramus, Savile, and the Renaissance Rediscovery of Mathematical History*, New Studies in the History of Science and Technology, Archimedes, #25, New York: Springer.

Goulding, Robert (2018), "Five Versions of Ramus's Geometry," in Margaret Meserve and Anthony Ossa-Richardson eds., *Et Amicorum: Essays on Renaissance Humanism and Philosophy in Honour of Jill Kraye*, 355–87, Leiden: Brill.

Grabiner, Judith (1998), "Some Disputes of Consequence: Maclaurin among the Molasses Barrels," *Social Studies of Science*, 28 (1): 139–68.

Grattan-Guinness, Ivor, ed. (1980 [2000]), *From the Calculus to Set Theory, 1630–1910*, London: Duckworth.

Grattan-Guinness, Ivor, ed. (1994), *Companion Encyclopedia of the History and Philosophy of the Mathematical Sciences*, vol. 1, London: Routledge.

Guicciardini, Niccolò (1999), *Reading the Principia. The Debate on Newton's Mathematical Methods for Natural Philosophy from 1687 to 1736*, Cambridge: Cambridge University Press.

Guicciardini, Niccolò (2009), *Isaac Newton on Mathematical Certainty and Method*, Cambridge, MA: MIT Press.

Hacking, Ian ([1975] 2006), *The Emergence of Probability*, Cambridge: Cambridge University Press.

Hägermann, Dieter, Wolfgang König, and Helmuth Schneider (2003), *Propyläen Technikgeschichte*, vol. *2: 1000 bis 1600*, Berlin: Propyläen-Verlag.

Hall, A. Rupert (1965), "Wren's Problem," *Notes and Records of the Royal Society of London*, 20 (2): 140–4.

Hall, A. Rupert (1980), *Philosophers at War: The Quarrel between Newton and Leibniz*, Cambridge: Cambridge University Press.

Harley, John Brian, and David Woodward, eds. (1987), *The History of Cartography, vol. 1: Cartography in Prehistoric, Ancient, and Medieval Europe and the Mediterranean*, Chicago, IL: University of Chicago Press.

Heeffer, Albrecht (2004), "*Récréations mathématiques* (1624): A Study on its Authorship, Sources and Influence," PhD thesis, University of Ghent.

Heeffer, Albrecht (2010), "The Reception of Ancient Indian Mathematics by Western Historians," in B. S. Yadav and Man Mohan eds., *Ancient Indian Leaps into Mathematics*, 135–52, Basel: Birkhäuser.

Hooykaas, Reyer (1958), *Humanisme, science et réforme: Pierre de La Ramée (1515–1572)*, Leiden: Brill.

Horiuchi, Annick ([1994] 2010), *Japanese Mathematics in the Edo Period (1600–1868): A Study of the Works of Seki Takakazu (?–1708) and Takebe Katahiro (1664–1739)*, Basel: Birkhäuser.

Horkheimer, Max, and Theodor W. Adorno ([1944] 2002), *Dialectic of Enlightenment*, ed. Gunzelin Schmid Noerr and trans. Edmund Jephcott, Stanford, CA: Stanford University Press.

Horwich, Paul, ed. (1993), *World Changes: Thomas Kuhn and the Nature of Science*, Cambridge, MA: MIT Press.

Hsia, Florence C. (2009), *Sojourners in a Strange Land: Jesuits and Their Scientific Missions in Late Imperial China*, Chicago, IL: University of Chicago Press.

Hughes, Barnabas (1993), "Robert Recorde and the First Published Equation," in Menso Folkerts and Jan Pieter Hogendijk eds., *Vestigia Mathematica: Studies in Medieval and Early Modern Mathematics in Honour of H.L.L. Busard*, 163–72, Amsterdam: Rodopi.

Huppert, Anne (2014), "Practical Mathematics in the Drawings of Baldassarre Peruzzi and Antonio da Sangallo the Younger," in Anthony Gerbino ed., *Geometrical*

Objects: Architecture and the Mathematical Sciences 1400–1800, 79–106, Cham: Springer.

Huygens, Chistiaan (1986), *The Pendulum Clock or Geometrical Demonstrations Concerning the Motion of Pendula as Applied to Clocks*, trans. Richard J. Blackwell, Ames, IA: Iowa State University Press.

Imhausen, Annette, and Volker Remmert (2006), "The Oration on the Dignity and the Usefulness of the Mathematical Sciences of Martinus Hortensius (Amsterdam, 1634): Text, Translation and Commentary," *History of Universities*, 21 (1): 71–150.

Jahnke, Hans Niels, ed. ([1999] 2003), *A History of Analysis*, Providence, RI/London: American Mathematical Society/London Mathematical Society.

Jami, Catherine (2011), *The Emperor's New Mathematics: Western Learning and Imperial Authority during the Kangxi Reign (1662–1722)*, Oxford: Oxford University Press.

Jesseph, Douglas M (1999), *Squaring the Circle: The War between Hobbes and Wallis*, Chicago, IL: University of Chicago Press.

Johnston, Steven (1991), "Mathematical Practitioners and Instruments in Elizabethan England," *Annals of Science*, 48 (4): 319–44.

Johnston, Steven (1994), "Making Mathematical Practice: Gentlemen, Practitioners and Artisans in Elizabethan England," PhD thesis, University of Cambridge.

Jullien, Vincent, ed. (2002), *Le calcul des longitudes*, Rennes: Presses Universitaires de Rennes.

Jullien, Vincent, ed. (2015), *Seventeenth-Century Indivisibles Revisited*, Cham: Birkhäuser.

Karp, Alexander, and Gert Schubring (2014), "Mathematics Education in Europe in the Premodern Times," in Alexander Karp and Gert Schubring eds., *Handbook on the History of Mathematics Education*, 129–52, Dordrecht: Springer.

Karp, Alexander, and Gert Schubring, eds. (2014), *Handbook on the History of Mathematics Education*, Dordrecht: Springer.

Karr Schmidt, Suzanne (2018), *Interactive and Sculptural Printmaking in the Renaissance*, Boston, MA: Brill.

Kepler, Johannes ([1604] 2000), *Optics: Paralipomena to Witelo & Optical Part of Astronomy*, trans. William H. Donahue, Santa Fe, NM: Green Lion Press.

Klaiber, Susan (2014), "Architecture and Mathematics in Early Modern Religious Orders," in Anthony Gerbino ed., *Geometrical Objects: Architecture and the Mathematical Sciences 1400–1800*, 137–64, Cham: Springer.

Knobloch, Eberhard (1994), "Mathematical Methods in Medieval and Renaissance Technology and Machines," in Ivor Grattan-Guinness ed., *Companion Encyclopedia of the History and Philosophy of the Mathematical Sciences*, vol. 1, 251–8, London: Routledge.

Koyré, Alexandre ([1939] 1966), *Études Galiléennes*, Paris: Hermann.

Koyré, Alexandre (1957), *From the Closed World to the Infinite Universe*, Baltimore, MD: Johns Hopkins University Press.

Krautheimer, Richard (1970), *Lorenzo Ghiberti*, vol. 1, Princeton, NJ: Princeton University Press.

Kuhn, Thomas (1957), *The Copernican Revolution: Planetary Astronomy in the Development of Western Thought*, Boston, MA: Harvard University Press.

Lardinois, André, Sophie Levie, Hans Hoeken, and Christoph Lüthy, eds. (2015), *Texts, Transmissions, Receptions: Modern Approaches to Narrative*, Radboud Studies in Humanities, #1, Leiden: Brill.

La Roche, Éstienne de, and Gilles Huguetan (1538), *L'Arismetique et geometrie*, Lyon: Gilles et Jacques Huguetan frères.

Leibniz, Gottfried W. (1684), "Nova methodus pro maximis et minimis, itemque tangentibus, quae nec fractas, nec irrationales quantitates moratur, et singulare pro illis calculi genus, *Acta Eruditorum*, III: 467–73.

Leibniz, Gottfried W. (1951), *Selections*, ed. and trans. Philip P. Wiener, New York: Scribner's.

Leibniz, Gottfried W. (1981), *New Essays on Human Understanding*, trans. Peter Remnant and Jonathan Bennett, Cambridge: Cambridge University Press.

Leibniz, Gottfried W. (1994), *Writings on China*, trans. Daniel J. Cook and Henry Rosemont, Chicago, IL: Open Court.

Leitão, Henrique, and Samuel Gessner (2014), "Euclid in Tiles: The Mathematical Azulejos of the Jesuit College in Coimbra," *Mathematische Semesterberichte*, 61: 1–5.

Long, Pamela O. (2011), *Artisan/Practitioners and the Rise of the New Sciences, 1400–1600*, Corvallis, OR: Oregon State University Press.

Long, Pamela O. (2015), "Trading Zones in Early Modern Europe," *Isis*, 106 (4): 840–7.

Long, Pamela O. (2018), *Engineering the Eternal City: Infrastructure, Topography, and the Culture of Knowledge in Late Sixteenth-Century Rome*, Chicago, IL: Chicago University Press.

Mancosu, Paolo (1996), *Philosophy of Mathematics and Mathematical Practice in the Seventeenth Century*, Oxford: Oxford University Press.

Marr, Alexander, ed. (2009), *The Worlds of Oronce Fine: Mathematics, Instruments and Print in Renaissance France*, Donnington: Shaun Tyas.

Marten, Bettina, Ulrich Reinisch, and Michael Korey, eds. (2012), *Festungsbau: Geometrie – Technologie – Sublimierung*, Berlin: Lukas Verlag für Kunst- und Geistesgeschichte.

McNeely, Ian F. (2009), "The Renaissance Academies between Science and the Humanities," *Configurations*, 17 (3): 227–58.

Melanchthon, Philip (1999), *Orations on Philosophy and Education*, ed. Sachiko Kusukawa and trans. Christine F. Salazar, Cambridge: Cambridge University Press.

Merrill, Elizabeth M. (2017), "Pocket-Size Architectural Notebooks and the Codification of Practical Knowledge," in Matteo Valleriani ed., *The Structures of Practical Knowledge*, 21–54, Cham: Springer.

Meserve, Margaret, and Anthony Ossa-Richardson, eds. (2018), *Et Amicorum: Essays on Renaissance Humanism and Philosophy in Honour of Jill Kraye*, 355–87, Leiden: Brill.

Møller Pedersen, Kirsti (1980 [2000]), "Techniques of the Calculus, 1630–1660," in Ivor Grattan-Guinness ed., *From the Calculus to Set Theory, 1630–1910: An Introductory History*, 10–48, London: Duckworth.

Moran, Bruce T. (1978), *Science at the Court of Hesse-Kassel: Informal Communication, Collaboration and the Role of the Prince-Practitioner in the Sixteenth Century*, Los Angeles, CA: University of California Press.

Morel, Thomas (2020), "De Re Geometrica: Writing, Drawing and Preaching Mathematics in Early Modern Mines," *Isis*, 111 (1): 22–45.

Morel, Thomas (2022), *Underground Mathematics. Craft Culture and Knowledge Production in Early Modern Europe*, Cambridge: Cambridge University Press.

Muffato, Martin (2018), *L'enseignement de l'arithmétique au féminin au XVII*[e] *siècle*. Available online: https://images.math.cnrs.fr/L-enseignement-de-l-arithmetique-au-feminin-au-XVIIe-siecle.html?lang=fr (accessed April 26, 2023).

Mukerji, Chandra (2009), *Impossible Engineering*, Princeton, NJ: Princeton University Press.

Needham, Joseph (1959), *Science and Civilisation in China: Mathematics and the Sciences of the Heavens and the Earth*, Cambridge: Cambridge University Press.

Newton, Isaac ([1967–81] 2008), *The Mathematical Papers of Isaac Newton*, ed. Derek T. Whiteside, 8 vols, Cambridge: Cambridge University Press.

Noyes, Ruth, ed. (2022), *Reassessing Epistemic Images in the Early Modern World*, Amsterdam: Amsterdam University Press.

Olson, Richard (1990), *Science Deified and Science Defied: The Historical Significance of Science in Western Culture*, vol. 2, Berkeley, CA: University of California Press.

Oosterhoff, Richard J. (2018), *University and Print in the Circle of Lefèvre d'Étaples*, Oxford: Oxford University Press.

Osler, Margaret J. (2010), *Reconfiguring the World: Nature, God, and Human Understanding from the Middle Ages to Early Modern Europe*, Baltimore, MD: Johns Hopkins University Press.

Parker, Robert H., and Basil S. Yamey, eds. (1994), *Accounting History: Some British Contributions*, Oxford: Clarendon Press.

Pascal, Blaise (1659), *Lettres de A. Dettonville contenant quelques-unes de ses inventions de géométrie*, Paris: Guillaume Desprez.

Pascal, Blaise (1910), *Thoughts: Letters. MIvor Works*, ed. and trans. Charles W. Eliot, William F. Trotter, Miriam L. Booth, and O. W. Wight, New York: Collier.

Pascal, Blaise (1968), *Pensées*, trans. A. J. Krailsheimer, London: Penguin Classics.

Petty, William ([1899] 1986), *The Economic Writings of Sir William Petty*, ed. C. H. Hull, 2 vols in 1, Fairfield, NJ: A.M. Kelley.

Pumfrey, Stephen, Paolo L. Rossi, and Maurice Slawinski, eds. (1991), *Science, Culture and Popular Belief in Renaissance Europe*, Manchester: Manchester University Press.

Rampling, Jennifer M. (2011), "The Elizabethan Mathematics of Everything: John Dee's 'Mathematicall Praeface' to Euclid's *Elements*," *Bulletin of the British Society for the History of Mathematics*, 26: 135–46.

Ramus, Petrus, and Audomarus Talaeus ([1599] 1969), *Collectaneae praefationes, epistolae, orationes*, Marburg edition. Reprint with an introduction by Walter J. Ong, Hildesheim, NY: Olms.

Reeves, Eileen (2008), *Galileo's Glassworks: The Telescope and the Mirror*, Cambridge, MA: Harvard University Press.

Remmert, Volker (2011), *Picturing the Scientific Revolution: Title Engravings in Early Modern Scientific Publications*, Philadelphia, PA: Saint Joseph's University Press.

Remmert, Volker (2016), "The Art of Garden and Landscape Design and the Mathematical Sciences in the Early Modern Period," in Hubertus Fischer, Volker R. Remmert, and Joachim Wolschke-Bulmahn eds., *Gardens, Knowledge, and the Sciences in the Early Modern Period*, Trends in the History of Science, 9–28, Basel: Birkhäuser.

Reynolds, Terry S. (1983), *Stronger than a Hundred Men: A History of the Vertical Water Wheel*, Baltimore, MD: Johns Hopkins University Press.

Robson, Eleanor, and Jacqueline Stedall, eds. ([2009] 2011), *The Oxford Handbook of the History of Mathematics*, Oxford: Oxford University Press.

Rose, Paul Lawrence (1975), *The Italian Renaissance of Mathematics: Studies on Humanists and Mathematicians from Petrarch to Galileo*, Geneva: Droz.

Saliba, George (2007), *Islamic Science and the Making of the European Renaissance*, Cambridge, MA: MIT Press.

Schubring, Gert (2005), *Conflicts between Generalization, Rigor, and Intuition: Number Concepts Underlying the Development of Analysis in 17th–19th-Century France*, New York: Springer.

Schüling, Hermann (1969), *Geschichte der axiomatischen Methode im 16. und beginnenden 17. Jahrhundert*, Hildesheim, NY: Olms.

Sergio, Emilio (2006), *Verità matematiche e forme della natura da Galileo a Newton*, Rome: Aracne editrice.

Simpkins, Diana M. (1966), "Early Editions of Euclid in England," *Annals of Science*, 22 (4): 225–49.

Skaarup, Bjørn Okholm (2015), "Applied Science in the Renaissance Art Academy," in Camilla Skovbjerg Paldam and Jacob Wamberg eds., *Art, Technology and Nature: Renaissance to Postmodernity*, 105–15, Farnham: Ashgate.

Skovbjerg Paldam, Camilla, and Jacob Wamberg, eds. (2015), *Art, Technology and Nature: Renaissance to Postmodernity*. Farnham: Ashgate.

Smith, A. Mark (2014), *From Sight to Light: The Passage from Ancient to Modern Optics*, Chicago, IL: University of Chicago Press.

Spinoza, Baruch (1883–4), *The Chief Works of Spinoza*, trans. Robert H. M. Elwes, 2 vols, London: George Bell.

Stedall, Jacqueline (2002), *A Discourse Concerning Algebra: English Algebra to 1685*, Oxford: Oxford University Press.

Stedall, Jacqueline (2008), *Mathematics Emerging: A Sourcebook 1540–1900*, Oxford: Oxford University Press.

Stephenson, Bruce (1987), *Kepler's Physical Astronomy*, Princeton, NJ: Princeton University Press.

Swerdlow, Noel M. (1993), "Science and Humanism in the Renaissance: Regiomontanus's Oration on the Dignity and Utility of the Mathematical Sciences," in Paul Horwich ed., *World Changes: Thomas Kuhn and the Nature of Science*, 131–68, Cambridge, MA: MIT Press.

Swerdlow, Noel M., and O. Neugebauer (1984), *Mathematical Astronomy in Copernicus's De Revolutionibus*, 2 vols, New York: Springer.

Tallemant des Réaux, Gédéon (1834), *Les Historiettes*. Mémoires pour servir à l'histoire du XVIIe siècle, publiés sur le manuscrit inédit et autographe par Messieurs Monmerqué, de Chateaugiron et Taschereau, vol. 1, Paris: A. Levavasseur.

Tallemant des Réaux, Gédéon (1834–5), *Historiettes*, vol. 1, Paris: A. Levavasseur.

Thomas, Keith (1987), "Numeracy in Early Modern England: The Prothero Lecture," *Transactions of the Royal Historical Society*, 37: 103–32.

Thomas-Stanford, Charles (1924), "Early Editions of Euclid's *Elements* 1482–1600," *The Library*, s4-V (1): 39–42.

Thrower, Norman J. W. (1996), *Maps & Civilization: Cartography in Culture and Society*, Chicago, IL: University of Chicago Press.

Torricelli, Evangelista (1644), *Opera Geometrica*, Florence: Amadoro Massa and Lorenzo de Landis.

Trenchant, Jean (1558), *L'Arithmétique. Departie en Troys Livres: Ensemble Un petit discours des changes. Avec L'art de calculer aux Getons*, Lyons: Jean Trenchant.

Turner, Anthony (2008), "'Not to Hurt of Trade': Guilds and Innovation in Horology and Precision Instrument Making," in Stephan R. Epstein and Maarten Prak eds., *Guilds, Innovation, and the European Economy, 1400–1800*, 264–87, Cambridge: Cambridge University Press.

Valleriani, Matteo, ed. (2017), *The Structures of Practical Knowledge*, Dordrecht: Springer.

van Damme, Stéphane, ed. (2015), *Histoire des sciences et des savoirs, vol. I: De la Renaissance aux Lumières*, Paris: Le Seuil.

Van Helden, Albert, Sven Dupré, Rob van Gent, and Huib Zuidervaart, eds. (2010), *The Origins of the Telescope*, Amsterdam: Royal Netherlands Academy of Arts and Sciences Press.

Vérin, Hélène (1993), *La gloire des ingénieurs: L'intelligence technique du XVIe au XVIIIe siècle*, Paris: Albin Michel.

Vierlingh, Andries (1920), *Tractaet van Dyckagie*, ed. Johannes de Hullu, 's-Gravenhage: Martinus Nijhoff.

Wardhaugh, Benjamin (2012a), *Poor Robin's Propheties: A Curious Almanac, and the Everyday Mathematics of Georgian Britain*, Oxford: Oxford University Press.

Wardhaugh, Benjamin, ed. (2012b), *A Wealth of Numbers: An Anthology of 500 Years of Popular Mathematics Writing*, Princeton, NJ: Princeton University Press.

Wess, Jane (2012), "Avoiding Arithmetic, or the Material Culture of Not Learning Mathematics," *Journal of the British Society for the History of Mathematics*, 27 (2): 82–106.

Williams, Jack (2011), *Robert Recorde: Tudor Polymath, Expositor and Practitioner of Computation*, London: Springer.

Willmoth, Francis (1993), *Sir Jonas Moore: Practical Mathematics and Restoration Science*, Woodbridge: Boydell.

Woodward, David, ed. (2007), *Cartography in the European Renaissance*, The History of Cartography, #3, Chicago, IL: University of Chicago Press.

Yadav, B. S., and Man Mohan, eds. (2010), *Ancient Indian Leaps into Mathematics*, Basel: Birkhäuser.

Yoder, Joella (1989), *Unrolling Time: Christiaan Huygens and the Mathematization of Nature*, Cambridge: Cambridge University Press.

Zemon Davis, Natalie (1958) "Mathematicians in the Sixteenth-Century French Academies: Some Further Evidence," *Renaissance News* 11: 3–10.

INDEX

Locators in *italics* refer to illustrations

abaci 39, 41, 42–3, 53
abbreviations 17–18
Abraham, Jean 46, 59
Academia Parisiensis 91
academies 28, 206, 208
  art 79
Accademia del Cimento 90
Accademia del Disegno 90
Accademia dei Lincei 89
Acta Eruditorum 91, 107
administration (administrators) 26, 198, 204
Aerarium Philosophiae Mathematicae 231
age of discovery 152–6
Agricola, Georgius 183, 190
Alberti, Leon Battista 81, 200
algebra 30, 127, 141, 143, 216
  between arithmetic and geometry 110–12
algorisms 42
algorithms, systematic 121
al-Khayyam, Omar 30
al-Khwārizmī, Muhammad ibn Musa 30, 42, 110
Allen, Elias 89
Almagestum novum 148, *149*
almanacs 26, 64, *65*
Ambassadors, The 159
Amsterdam, University of 20
analysis 31
anamorphoses 159
Anatomy of the World 5–6
Ancients 98–104
annuities 174
Apian, Peter 71, 152
Apian, Philipp 73
Apollonius 98, 106
  problems of 92
apprenticeships 50
Archimedes 98–9, 210, 222, 225
architects 53
  see also artist-engineers
architecture 9, 76, 80, 85, 198–205
areas 103
Aristotle 4, 125, 128, 210
arithmetic 207, 218
Arithmetica 44, *45*
arithmetic (personification) 219
  see also personifications
arithmetic, political 172–4
arithmétique au miroir, L' 63
armies (military) 25, 61
Arnauld, Antoine 142
arsenals 15
Ars magna sciendi sive combinatoria 136, *137*
artificial perspective 156–9
artillery 61, 80
artisan mathematicians 83

artisans 80, 158–9, 180, 182, 197, 198, 200–1, 203, 205
artists 53, 158–9
Ascham, Roger 49–50
astrolabes 68, 72, 226
astrology 64, 218
astronomy (astronomers) 9, 74–6, 144
  representing mathematics 216, 218, 226
  World, describing and understanding 151, 174
audiences 10, 56, 64
authority 212, 216
Averroes 128, 130

Bachet de Méziriac, Claude Gaspard 62
Bacon, Francis 7–8, 205, 231
Bacon, Nicholas 214
Bacon, Roger 214
bakers 197
ballistics 29, 92, 163, 167
banking systems 3, 26
Barbaro, Daniele 76
Barozzi, Francesco 131, 210
barrels 194
  see also gauging
Barrême, François 59, 62
Barrow, Isaac 33, 99, 115–17, 119, 141–2, 176
bartering 60
Beaugrand, Jean de 113
Beemster Lake 188
Bélidor, Bernard Forest de 186
Bell, Johann Adam Schall von 77, 144
Benedetti, Giovanni Battista 75, 164
Bernoulli, Jakob (Jacob) 35, 120, 156
Bernoulli, Johann 32, 35, 120, 142
Bertoloni Meli, Domenico 8
Bessarion, Cardinal 10
Besson, Jacques 183
Bettini, Mario 231
Billingsley, Henry 215
Blaeu, Willem 84
boarding schools 52
Bollaert, Roeland 73
Bombelli, Rafael 31, 143
booklets 25, 28, 72, 87–9, 231
Borelli, Giovanni Alfonso 21
Bos, Henk 105
Botteghe d'ab(b)aco 53
Bourne, William 214
Boyle, Robert 3, 131
Brahe, Tycho 151
Bradley, Humphrey 189
Brentjes, Sonja 207
Briggs, Henry 93, 114, 193
Brouncker, William 90
Brunelleschi, Filippo 80–1, 198, 199–200, 201
Bürgi, Jost 75, 226
business contracts 173

Caballa, Christian 136
cabinets of curiosity 90
calculators 47
calculus 34, 117–21, 127, 136
calendars 11, 64, 78, 80
  Chinese 144
  Gregorian 77
  Julian 76
camera obcura 160, 214
Campanella, Tommaso 131
Campanus of Novara 97
Canal du Midi 199, 203–5
canals 203
capitalism 126, 194
Cardano, Gerolamo (Girolamo) 30–1, 86, 88, *109*, 110
careers, mathematical 77, 93
carpenters 193, 197
cartography 84, 85, 87, 172, 174
  age of discovery 152–6
  see also maps (mapmaking)
casks 194, 197
  see also gauging
Cataldi, Pietro Antonio 79
causality 164
Cavalieri, Bonaventura 33, 113, 119, 168
certainty 212
  mathematical 128–31
  in seventeenth century 131–3
  see also worldviews
Ceulen, Ludolph van 81
Chales, Jean-François Milliet de 27, 225
chance, games of 172–4
characteristic triangle *117*, 121
charity schools 51
charts 192–8, 205–6
  geometric planning and architecture 198–205

charts, nautical 152
China 144–5
Chrisomalis, Stephen 46
Chronica regia coloniensis 209
Chuquet, Nicolas 54
churches 74–9
circles 135–6
circumnavigations 3
cities 79–84
Clavius, Christopher 77, *78*, 98, 131, 211, 216
clocks 168
  astronomical 226
  see also timekeeping
clocks, pendulum 170
Colbert, Controller-General 203–4
Collado, Luis 25
Collège Royal, Paris 79
colleges 14, 48, 74–9
Collins, John 118, 155
colonialism 198
combinatorics 136
Commandino, Federico 28–9, 72, 98, 159, 216
commerce 58
commodities 3
communication 58
companies, chartered 198
compasses 102, 152
  makers 82
  mining 191
competitions 201
  see also contests
complainte de François Garin, La 49
complex numbers 31
composition of movements 116
compound interest 194
computations 140
conics 4, 102
Conics 98
Connaissance des temps 91
Constantinople 2
contests 89–92
Copernican Revolution 127
Copernicus, Nicolaus 6, 27, 71, 125, 148–52
copying 17
corporations 79–84
correspondence 89–92
Cort, Cornelis (Pieter Cornelisz) 188, 218–19
Cosa, Juan de la 153
cosine, hyperbolic 32
cosmography 68, 70
cosmology 151–2
counting boards 57
counting mats 43
courts 74–9, 225
craftsmen (crafts) 14, 197, 206
crises 177, 216
cubic rod 197
curved locks 205
curves 32, 34, 102, 104, 118, 171
cycloids 32, 169
Cyclometrica elementa duo 216, *217*

daily life 55–65
Danti, Egnatio (Ignazio) 77, 79, 203
Darling, John 57
da Sangallo, Antonio the younger 76
data collection 57
da Vinci, Leonardo 80, 182, 184–5, 201, 203
debates 131, 133
  new mathematics 141–2
decimal fractions 58
decimal place-value system 42
decision-making 174
Dee, John 153, 213
  everyday numeracy 49, 56, 61
  practice and profession 67, 69–70, 72, 84
De geometria recondita et analysi indivisibilium atque infinitorum 120
Della Porta, Giambattista (Giovanni Battista) 160, 214, 229
Della trasportatione dell'obelisco vaticano 201
del Monte, Guidobaldo 81, 159
De revolutionibus orbium coelestium 5
de Sacrobosco, John 10
Desaguliers, John 184
Desargues, Girard 29–30, 91, 112–13, 159
Descartes, René 6, 31, 34, 91, 177
  inventing mathematics 98, 101–9, 111
  worldviews 127, 131–5, 138–9
development, economic 188
diagrams 16–17, 215
differential calculus 121, 40
Digges, Thomas 86, 214
di Giorgio, Francesco 201

Diophantus of Alexandria 110
Dioptrice 163
Discours de la méthode 134
Discourses and Mathematical Demonstrations Relating to Two New Sciences 164–5, 167
dissemination 6, 62
domes 200
Donne, John 5–6
double-entry bookkeeping 3
double infinity 139
Dou, Jan Pietersz 189
drainage techniques 187–90
Dürer, Albrecht 12, 25, 71, 182
Dutch East India Company (VOC) 84
Dutch War of Independence 188
dykes 187–90

East India Company (EIC) 84
economies 126
  development 188
  political 173
education 12, 14, 28, 77
  everyday numeracy 48, 50–2
  universities 218
  worldviews 132, 143
Eighty Years War 26
Elements 4, 29, 97, 98, 127, 215
elites 4, 6, 13, 26, 173, 218
  everyday numeracy 48, 50–2
emancipation movements 51
émigrés 3–4
employment 28
energy sources 182
engineering 9, 77, 201
engineers 181, 182, 184
  see also artists-engineers
entertainments 28, 229, 231
Enumeration of Lines of the Third Order 111
equal signs 216
equations 30–1, 101, 106
Erasmus, Desiderius 13, 207
Étaples, Jacques Lefèvre d' 216
Euclid 4, 135, 210, 222, 225
Euclidis quae supersunt omnia 224
Eudemus of Rhodes 210
Eurocentrism 128
exchange rates, currency 60
excises 80
experimental philosophers 131

Faulhaber, Johannes 9, 81, 219
Favino, Federica 9
Fenlands 189
Fermat, Pierre de 91, 106, 111, 115, 173
finance 53, 193–4
Fine (Finé), Oronce 22, 49, 216
finger-counting 38–9
Fire of London (1666) 85
firing tables 29
first and last ratios 140
Flanders 26
Florence, Dome of 198
fluxional calculus 140
Fontana, Domenico 199, 201, 202–3
Fontana, Niccolò (Tartaglia) 29
  see also Tartaglia, Niccolò
Fontenelle, Bernard de 17
fortifications 25, 85, 229
  see also military
Foucault, Michel 126
fractions 59
French Revolution 26
Frisius, Gemma 49, 55, 68, *69*, 73, 187
From the Closed World to the Infinite Universe 5
fundamental theorem of calculus 121

Galilei, Galileo (Galileo) 6–8, 15–16, 21, 27
  invention of mathematics 104
  practice and profession 75
  representing mathematics 231
  technology 185
  World, describing and understanding 163–8, 176
  worldviews 125, 128
games of chance 172–4
gardens 231
Gassendi, Pierre 131
gauging 85, 194
gauging rods 195, *196*, 197
Gellibrand, Henry 93
gelosia multiplication method *43*
general binomial theorem 121
General Heads for the Natural History of a Country 3

geographical spread (north) 92–4
geography 154
Geography 22
geometrical exactness 106
Geometrical Lectures 176
Géométrie 101
geometry 14
  algebraization of 141
  Greek 100–1
  representing mathematics 207, 216, 218, 226
Geometry 104–5, 111, 118
Ghetaldi, Marinus 101
girls 51–2, 57
globes 68, 70
gnomonics 85
Gogava, Antonio 68–70
golden rule 59–60
Golius, Jacob 100
Grassi, Orazio 76
Gregory, James 118, 155
Gregory of Saint Vincent 33
Gresham College 93
guilds 82
gunnery 85, 163
  see also artillery
Gunter, Edmund 89

Halley, Edmond 156
Harriot, Thomas 93, 112
Hartmann, Georg 71
Hartwell, Robert 45
Heuraet, Hendrik van 118
Hevelius, Johannes 13, 91
Histoire et Mémoires de l'Académie des Sciences 107
Hobbes, Thomas 141–2, 177
Hooke, Robert 85, 94
horologium oscillatorium 171
horoscopes 9
Hortensius, Martinus 20, 227
Hudde, Johannes 84, 118
humanism 4, 48
Huygens, Christiaan 32, 91, 126, 186
  World, describing and understanding 168–9, 171–3, 175–6
Huygens, Constantin 108
hydraulics 80, 184, 186, 204
hyperbolic cosine 32

Ibn al-Haytham 156
Iconologia 222, *223*
imagery 218
images 16–17
imago 162
imitatio 163
indeterminate coefficients 111
Indian calculus 143
indivisibles 112–17
Industrial Revolution 180, 186
infinite series 33–4
infinitesimal calculus 35, 126, 177
infinitesimals 112–17, 168, 169
infinity 138–141
information dissemination 201
information storage 193
Ingenieurs-Schul 220
innovation 4, 107
institutions 9, 12, 28, 79, 83–4, 131
instrument making 9, 82–3
instruments 22, 87–9, 226
  paper 17
  scientific 5, 15
insurance systems 3, 173
integral calculus 121, 195
interest, simple and compound 60, 194
Introduction for to lerne to Reckon with the Pen, or with the counters, An 44

Japanese mathematics 143
Jesuits 6, 11, 14, 21–2, 27, 76–7
Josephus, Flavius 208
journals 6, 89–92, 107
Journal des sçavans 91

Kepler, Johannes 6, 22, 33, 73–5, 126
  representing mathematics 215, 231
  World, describing and understanding 147, 150–1, 156, 159–63
Kilian, Lukas 218
kinematics 147
Kircher, Athanasius 136
knowledge 4–5, 6, 134, 139, 222
  astronomical 144
  craft 180
  hierarchy of 221, 227
  mathematical 212
  see also practical mathematics

Koopmann, Elisabeth 13
Koyré, Alexandre 126, 128, 138, 151

land reclamation 187–8, 189
  see also polders
languages 52, 53, 55
  symbolic 126, 142
Lansbergen, Philipp 22
la Roche, Estienne de 61
la Salle, Jean-Baptiste de 51
lattitude 152
Lavanha, João Baptista 79
Leibniz, Gottfried Wilhelm 34–5, 191
  inventing mathematics 97, 115–21
  worldviews 127, 136, 138, 140–2, 144–5
Leiden University 81, 93–4
lengths 103
letters *see* correspondence
l'Hôpital, Guillaume de 120
Liber Mechanicorum 202
life insurance 57
linear perspective 157
livre de chifres et de gects, Le 44
Llull, Ramon 136
locks 204–5
logarithmic spirals 155–6
logarthims 193
longitude 152
Longobardo, Nicolò 78–9
loxodromes 153–5
Luther, Martin 2

Machine de Marly 186
machines 199, 206
Maestlin, Michael 73
Magia naturalis 214
Magia naturlis 229
magnetism 176
Malebranche, Nicolas 127
Manetti, Antonio 200
Mans, Jacques Peletier du 55
manuals 152
Manutius, Aldus 4
maps (mapmaking) 9, 25, 68, 147, 187–92
  see also cartography
Mariotte, Edme 186
maritime 83–4
Martini, Francesco di Georgio 185
Mathematial Principles of Natural Philosophy (Principia) 95, 98
Mathematical Collection 4
Mathematical Dictionary 227, 228
mathematical inventions 97–8
  algebra between arithmetic and geometry 110–12
  dialectics between Ancients and Moderns 98–104
  from indivisibles to infinitesimals 112–17
  invention of calculus by Newton and Leibniz 117–21
  solving problems 104–7
  writing 107–9
Mathematical Lectures 176
Mathematical Papers of Isaac Newton 119
mathematical physics 120
mathematical practices, various settings 11–16
mathematical sciences 175–6
  histories 208–11
  utility 226–3
mathematical sciences, early modern 8–11
mathematization 5, 7–8, 126, 181, 206
  forms of 19–26
  nature 174–7
Maurice of Nassau 133
Maurolico, Francesco 28–9, 72–3
mensuration 22
measurements 80
measuring units 206
mechanics 20, 35, 203–4
  World, describing and understanding 147, 164, 174, 176
media, printed 28
  see also printing
Medici family 75
Medina, Pedro de 84
Melanchthon, Philip 13, 51, 70, 74, 210–11
Mellema, Jean 57
Mercator, Gerard 22, 68, 70, 73, 90, 154
Mercator, Nikolaus (Nicolaus) 118, 155
merchants 9, 42, 43, 53, 56, 197

merchants' rule 59
Merrill, Elizabeth 201
Mersenne, Marin 6, 91, 107–8, 133, 168, 176
metal alloys 61
methods 28–35
metrology 80
Méziriac, Claude-Gaspar Bachet de 231
Milan, Dome of 201
military 25, 61
mills *see* waterwheels
mining (miners) 15, 25, 183, 190–2, 206
Mirica, Gasper à 68
missionaries 7, 143
models 181, 199, 200, 202, 215
modernity 128
Moderns 98–104
Modern Science 127
molenmeester 184
Moore, Jonas 81–2, 189–90
morality (moral training) 51
Morland, Sir Samuel 47
mortality 25, 174
  see also insurance systems
motions 34, 163–72
Müller, Johannes (Regiomontanus) *see* Regiomontanus, Johannes
Münster, Sebastian 153
Muscovy Company 84
music 20–1, 218

Napier (Neper), John 46, 114, 193
Napier's rods (bones) 46–7, *48*
national wealth 172
natural philosophy 27, 75, 104
  World, describing and understanding 172, 176–7
  worldviews 125–6, 151–2
nature 20, 125–6, 139–40, 147
  early modern mathematization 174–7
navigation (navigators) 22, 152, 197
  practice and profession 84–5, 92–3
  representing mathematics 207, 227
Needham, Joseph 143
negative coordinates 112
negative numbers 31
Netherlands 53
New Systeme of the Mathematicks 229, *230*
Newton, Isaac 17, 34, 95, 231
  World, describing and understanding 155, 177
  worldviews 126, 140–2
  see also mathematical inventions
Niceron, Jean-François 159
nocturnals 22, *23–4*
Norwood, Richard 155
Nova methodus pro maximis et minimis itemque tangentibus 119, *120*
Nova Scientia 221
numbering trees 39–41
numeracy, everyday 37–8
  acquiring skills 48–55
  daily life 55–65
  reckoning systems 38–8
Nunes, Pedro 83–4, 88, 153
Nuremberg 53, 82
  1530 70–2

obelisks 202
objects, new 28–35
observatories 13, 72–4
Öder, Mattias 187
Oldenburg, Henry 6
Oosterhoff, Richard J. 11
optics 20, 21, 156–9, 174, 176, 214
  Kepler 159–63
Oughtred, William 89, 94
Oxford University 93–4, 211
Ozanam, Jacques 227, 229

painters 157, 163
paper technologies 192–8
Pappus of Alexandria 99–100, 102, 107
Parent, Antoine 186
Pascal, Blaise 32, 47, 91
  inventing mathematics 109, 113, 115
  World, describing and understanding 168, 173, 175
  worldviews 133, 139–40
patrons (patronage) 11, 28, 74, 77, 79
pendulum clocks 168, *170*, 171
pensions 90
Pereira, Benito 131
periodicals 107
personifications 218–19, 221–2, 229, 231, 233

perspective 76
  natural and artificial 156–9
perspective drawing 85
Petreius, Johannes 73
Petty, William 52, 172, 173, 174
Peucer, Casper 74
Philosophiae naturalis principia mathematica 1, 5
Philosophical Transactions 6, 91
Philosophical transactions of the Royal Society 107
Philosophiœ naturalis principia mathematica 140
physico-mathematics 176, 177
physics 20
Piccolomini, Alessandro 128, 130–1, 133
pictorial perspectives 156–9
picturae 162–3
Piero della Francesca 157–8
Pigafetta, Filippo 202
pilots, ship 152
Planer, Martin 184
planets 130, 148, *149*, 150
planning, geometric 198–205
Plantin, Christophe 73
Plato 4, 125–7, 210
pneumatics 176
polders 184, 187–90, 206
political arithmetic 172–4
politics 201
population growth 172
populations 57
portolans 152
Portugal 59
power 126
practical mathematics 174
  everyday numeracy 49, 53–5, 62
  see also practice and profession; technology
practice and profession 94–5
  mathematical activity in time and place 67–72
  practice and locus 72–84
  principles, tools, and trends 84–94
Practijk des Landmetens 189
practitioners, field 56
practitioners, mathematical 83, 180, 182, 206
prediction 150
Prestet, Jean 142
prestige 28, 153, 212
principles 84–9
Principles of Philosophy 138
printed media 28
printing 3, 4, 6, 11
  practice and profession 70
  representing mathematics 211
  technology 193
printing (print technology) 64, 72–4
probability, mathematical 173
probability theory 25, 174
Problèmes plaisans et délectables 231
problem-solving 15, 92, 101–3, 104–7
Proclus 128, 130–1, 210, 212, 213
Proclus commentary on Euclid 129
prognostications 64
projections 87, 153–4
projects 198
property 190–1
proportions 61, 103–4, 127
  theory of 140
prosperity 227, 229
Ptolemy 150
public accounts 43
publicity 203
publishing (publications) 64, 76, 193

quadrants 115
quantification 126, 173
quarto y vintena rule 59
questions, changing 28–35

Ramus, Pierre (Peter) 55, 75, 133, 205, 211, 212–13, 215–16
Ratdolt, Erhard 73, 215
Rathborne, Aaron 85
ratios 103
  first and last 34, 140
Ready Reckoners 46
reckoning systems 38–48
reckonmasters 53, 195
recognition 93
  social 206
Recorde, Robert 216
  everyday numeracy 39, 41, 43, 49
  practice and profession 84, 88
recreations 231
reflexivity 144

refraction 160–1
Regiomontanus, Johannes 10–11, 73, 89, 210
regulations 82
Reinhold, Erasmus 74
Reisch, Gregor 44
relationships 127
religion 51, 159, 172
Re metallica, De 190
  appearance of textbooks 215–18
  cultural and epsitemological status of mathematics 211–15
  histories of mathematical sciences 208–11
  representation, mathematical 207–8
  utility of mathematical sciences 226–33
  visual representations 218–26
resources 16–19
revolutionibus orbium coelestium, De 125
Reynolds, John 56–7, 58
Rheticus, Georg Joachim 70–1, 74
rhumb lines 152, 153, 155
rhymes, children's 57
Ricci, Matteo 7, 77, 143
Ricci, Ostilio 79
Riccioli, Giovanni Battista 148, 150, 168
Ricquet, Pierre-Paul 199, 203–5
riddles 62
Riese, Adam 43, 53, 61–2
Ripa, Cesare 222
Roberval, Gilles Personne de 33, 91, 114, 115–16, 168
rods, gauging *see* gauging rods
Roomen, Adriaan van 92, 218
Royal Society of London 6, 90, 131
Rudolph II, Emperor 74–5
rule of fellowship 60
rule of proportion 59
rule of three 59, 60
rulers 102
rules 134, 181

Saggiatore, Il 231, 232
sailors 212
Santcliment, Francesc 54
Savary, Jacques 60
Savile, Henry 211, 212
Savorgnan, Giulio 198
Scaliger, Joseph Justus 216, 218
Schöner, Andreas 27
Schöner, Johann 70–1
schools 9, 50, 152
  see also charity schools
Schooten, Frans van 141
Schott, Gasper 9, 225
sciences 147
Scientia nova 163
Scientific Revolution 4–8, 28, 124–5, 142
scientization 7
semi-circles 191
Sems, Johan 189
seven liberal arts 218
Sévigné, Madame de 57
skepticism 133
skills acquisition 48–55
slaves 172
Smeaton, John 186
Snell, Willebrord 153
social status 27, 28
societies 89–92
soroban 42
speculative mathematics 85–7
speed 164–5
Spinoza, Baruch 127, 135–6
spirals 102
  logarithmic 155–6
Stampioen, Jan 92
standardization 191, 193, 197, 206
standards 134
statics 20, 203
  see also mechanics
statistics 25
status 153, 159, 227
  social (cultural) and epistemological 26–8, 211–15
Stevin, Simon 16, 93, 111, 141
  everyday numeracy 58, 62
  technology 182, 189, 191, 194
  World, describing and understanding 159, 175
Stöffler, Johannes 72
strings, knotted 46
suanpan 41
subterranean geometry 190
subunits 58, 59
surveying (surveyors) 9, 22, 25, 57
  practice and profession 77, 80, 85, 92

representing mathematics 207, 226
technology 187, 189
underground 190–2
Surveyor, The 85, *86*
Swift, Jonathan 173
Syderum Secreta 64
symbolic algebra 141
symbolic writing 141
symbolism (symbols) 17–18, 205
inventing mathematics 99, 101, 108–9
worldviews 126, 136

tables 11, 22, 181, 192–8
Tacquet, André 113
Tacuinus, Joannes 215
Tangencies 106
tangents 115
Tartaglia, Niccolò 29, 61, 163–4, 222
practice and profession 75, 80, 91–2
taxes (taxation) 26, 57, 59, 172, 173, 195
technology 179–82
paper 192–8
seeing through stones 190–2
taming the seas 187–90
waterwheels 182–6
telescopes 163, 214
textbooks 13, 28, 85, 189, 192, 215–18
see also numeracy, everyday
theology 141
three R's 49, 50
timekeeping 9, 22, 25
time telling 80
tools 28–35, 84–9
Torricelli, Evangelista 33, 114, 168
Toscanelli, Paolo dal Pozzo 80
trade 26, 53, 92, 207, 227, 229
books 56
manufacturing 83
policies 172
tradesmen 212
trading companies 14
traditions 4
training, vocational 50
trajectories 29
transcendental curves 102
translations 16
Trenchant, Jean 45, 56, 194
truth 134
tuition, private 50
Turner, Anthony 83
Tycho, Brahe 73, 74–5
Typographia Medicea 7

underground surveying 190–2
universalism 144
universities 12–13, 72–4
see also representation, mathematical
University of Amsterdam 20
University of Leiden 81, 93–4
University of Oxford 93–4, 211
urban life, quality of 3
Ursuline schools 52

variable quantities 118
Vatican obelisk 199, 201
Vauban, Sébastien le Prestre de 205
venture capitalists 198
Verbiest, Ferdinand 78
Vierlingh, Andries 187
Viète, François 31, 86–7
inventing mathematics 101, 103, 108–9
worldviews 133, 141
Vigenère, Blaise de 54
visual representations 218–26
Vlacq, Adriaan 191, 193
vocational training 50
Voigtel, Nicolas 191
volvelles 17, 88
voyages of discovery 3, 5

Wallis, John 33–4, 50, 91, 141–3
inventing mathematics 114–15, 118
World, describing and understanding 155, 168
Walther, Bernhard 12, 73
warfare 61, 80, 163, 182, 227
see also military
water levels 191
watermills *see* waterwheels
waterwheels 182–6
weights 147
Widmann, Johnannes 193–4
Wiles, Andrew 111
William IV of Hessen-Kassel 226
Witelo (Vitello) of Silesia 156

Witt, Johann de 173–4
women 46, 51, 57
World, describing and understanding 147–8
  cartography in age of discovery 152–6
  Copernican astronomy 148–52
  early modern mathematization of nature 174–7
  Kepler's eye and geometrical optics 159–63
  mathematics of motion 163–72
  natural and artificial perspective 156–9
  political arithmetic and games of chance 172–4
worldviews 123–4
  certainty in seventeenth century 131–3
  debating new mathematics 141–2
  on the infinite 138–41
  in the manner of the geometers 134–8
  mathematical certainty 128–31
  Plato's revenge 125–7
  strange encounters 142–5
Wren, Christopher 32, 85, 90, 92, 94
Wright, Edward 93, 154–5
writing, symbolic 141

Zamberti, Bartolomeo 215
zero (nothing) 42, 44
Zilsel, Edgard 180